SACRED COWS, SACRED PLACES

SACRED COWS, SACRED PLACES

Origins and Survivals of Animal Homes in India

DERYCK O. LODRICK

UNIVERSITY OF CALIFORNIA PRESS
Berkeley Los Angeles London

University of California Press
Berkeley and Los Angeles, California

University of California Press, Ltd.
London, England

Library of Congress Cataloging in Publication Data

Lodrick Deryck O.
Sacred cows, sacred places.

Bibliography: p. 292
Includes index.
1. Cows (in religion, folklore, etc.) 2. Cattle—India. 3. Animals, Treatment of—India—Societies, etc. 4. Ahiṃsā. I. Title.
BL1215.C7L63 294.5'7 80-51240
ISBN 0-520-04109-7

Printed in the United States of America

For
Nana and John

Ahiṃsā parāmo dharma
Ahiṃsā is the greatest of religions

—Jaina Teaching

CONTENTS

PREFACE

Many thousands of miles and several years separate the plains and deserts of northern India, where this study was carried out, from the shores of Lake Mendota in Wisconsin, where I first heard of those unusual institutions in which useless cattle and other sick and disabled animals are afforded food and shelter until they die from natural causes rather than at the hands of man. A survey of literary sources revealed that, despite the voluminous writings available on India and on things Indian, animal homes are scarcely mentioned. They are noted as existing possibly as early as the third century B.C., are alluded to in scattered references since that time, but have received close attention only since the early decades of this century and then mainly from the perspective of their economic potential. I thus set out to undertake as comprehensive a study as possible of the cultural origins, historical development, and modern situation of animal homes in India.

Preliminary library work for the study was completed at the University of California at Berkeley, the India Institute at Oxford, and the India Office Library in London. My investigations continued in India where, as part of a doctoral program, I spent from January 1974 to September 1975 conducting field research supported by a Research Fellowship awarded by the American Institute of Indian Studies. During my stay in India, I was affiliated with Banaras Hindu University, and the Head of Geography, Professor R. L. Singh, extended to me full use of the facilities of his department. Dr. Kashi Nath Singh, also of Banaras Hindu University, Dr. A. B. Mukerji of the University of the Punjab at Chandigarh, and Dr. Harjivan Suthar of Gujarat University in Ahmedabad took considerable interest in my work, provided much useful information, and helped me to clarify and articulate many of my ideas.

Of the many other people who contributed to the successful completion of this study, only a few can be named here. Śrivatsa Goswami of Vrindavan was both companion and guide on my intellectual pilgrimage to Vraj; he introduced me to the complexities of Vaishnavism, and he continues to be an unfailing source of information on matters relating to the cult of Krishna. Ram Pal Agarwal of Patna and Anshuman Deva of Varānasi provided valuable assistance. But most of all, my appreciation goes to the late Dr. John Thompson-Wells, F.R.C.S., F.A.C.S., F.A.C.P., and his family at the Madar Hospital near Ajmer in Rajasthan. Their home was always open to me, and without this refuge in a country that is at one and the same time fascinating and extremely frustrating, it is unlikely that I could have accomplished all that I did in terms of fieldwork in India. Professors Frederick Simoons, Stephen Jett, and Wolfram Eberhard read the manuscript and suggested improvements. The illustrations are by Gene Christman, and the maps are by Marlin Dulay. Unless indicated otherwise, the photographs were taken by the author.

Foreign terms that appear frequently in the text, and proper names, such as Krishna, are presented in their common English forms, that is, *goshala* instead of *gośālā*, and Krishna instead of Kṛṣṇa. The system of transliteration of other Sanskrit, Hindi, and Gujarati words follows that set out by the Library of Congress. Foreign words are italicized the first time they appear in the text and thereafter only when the context requires it. A glossary of the more important terms is provided following the notes.

Finally, I would like to acknowledge my colleagues, students, and friends who, during the course of this study, have uncomplainingly suffered through repeated and enthusiastic lectures on animal homes, sacred cows, and sundry related topics.

San Francisco
November 1979 — D. O. L.

1

Man, Culture, and Animals in India

Ever since the early fourteenth century, when Marco Polo first astounded medieval Europe with the account of his epic journeys across Asia, European travelers to India have been intrigued by the many strange habits and customs they found there. In fact, so taken were they with the "marvyles of Inde" that almost three centuries later Ralph Fitch, one of the first Englishmen ever to visit India, could still write with some degree of incredulity:

> They have a very strange order among them, they worship a Cow, and esteeme much of the Cowes dung to paint the walls of their houses. They will kill nothing not so much as a Louse: for they hold it a sinne to kill any thing. They eate no flesh, but live by Roots, and Rice, and Milke. . . . In Cambaia they will kill nothing, nor have any thing killed: in the Towne they have Hospitals to keepe lame Dogs and Cats, and for Birds. They will give meate to the Ants.[1]

This interest in what, to the western mind, seem rather unusual practices has by no means diminished with the passing of years. Even today, some seven hundred years after Marco Polo noted that the people of Maabar worshipped the ox, refused to eat meat, and were reluctant to take life in any form,[2] these very traits lie at the heart of the debate on the sacred cow of India which is currently taking place among scholars from both India and the West.

Cow worship, the ritual use of dung, the philosophical concept of *ahiṃsā* (noninjury to sentient creatures), and the avoidance of meat—all characteristics noted by Fitch—have attracted the attention of students of culture and of India, yet the animal hospitals (*pinjrapoles*) and related institutions which serve as refuges for cattle (*goshalas*) have been strangely neglected. Some studies have been made by Indian writers in recent years, but these have generally been limited in scope and have often been undertaken by government officials concerned with resource evaluation, with a view to incorporating the institutions into post-Independence economic planning.[3] In the literature of the western world, however, animal homes have been dismissed as among the more obscure aspects of the cattle complex and as such worthy of mention only in passing. Thus, despite their uniqueness, there exists no comprehensive treatment of goshalas and pinjrapoles as elements in the cattle complex of India and as intriguing expressions of the interaction between man and animals in a traditional, nonwestern setting—a situation that I have set out to correct.

Although the study of man and animals forms a persistent theme in the literature of cultural geography and related disciplines, few scholars have concerned themselves with the case of India.[4] Yet perhaps nowhere else in the world do man and animals come together in such a rich and diversified cultural environment. Attitudes and behavior toward animals are so interwoven with the intricate fabric of Indian society that, as geographers working in this area have discovered, it becomes difficult to understand them without reference to their social, cultural, and historical milieu. Frederick Simoons, for example, has repeatedly found it necessary to invoke a variety of factors to explain patterns of behavior associated with animals in India. In his study of the mithan (*Bos [Bibos] frontalis*), a freeranging bovine domesticate of the northeastern hills, he concludes that strictly economic considerations were secondary in the husbandry complex to concerns of sacrifice and fertility, and that indeed the latter were most likely the driving force behind the domestication of the animal.[5] Simoons's studies of food habits in South Asia also illustrate the complex nature of the problem. He suggests, for instance, that the most important restrictions on the consumption of foods of animal origin derive from religions such as Hinduism and Buddhism which base their teachings on belief in the sanctity of all life. Thus vegetarianism in India, the rejection of beef by caste Hindus, and similar food avoidances can ultimately be traced to the ahiṃsā concept, though other socio-religious factors need to be considered in understanding how these traits evolved and how they survive in India today.[6] The ahiṃsā concept also helps explain why certain groups reject milk as a food source, but this is only one of a complex of elements. Simoons demonstrates that patterns of milk use and milk avoidance in South Asia reflect a wide range of ecological, physiological, and cultural factors.[7] Similarly, though he sees fish avoidance in India as being influenced by three main factors—the view that fish are unclean, the identification of fish with particular deities, and the ahiṃsā concept—Simoons argues that "one must take into account a multiplicity of factors, historical, socio-cultural and ecological, if one is adequately to understand the role fish and other potential foods play in human diet."[8]

Other studies support the view that in South Asia, unlike the West where domesticated herd animals are nothing more than a food source, behavior toward such animals is in part economic but always much more than that. An understanding of the entire physical, cultural, and historical context is thus essential to any study of India's animal husbandry complex. Robert Hoffpauir, for instance, in his study of the water buffalo (*Bubalus bubalus*) in India, finds that the low religious status of the animal and its "violability" in a society preeminently concerned with the inviolability of the zebu (*Bos indicus*) has significant economic ramifications. Not only is the buffalo an important sacrificial animal, but the absence of any religiously derived restrictions on husbandry practices has contributed to its position as the major milk producer of India.[9] Similarly, in his work with the yak (*Bos [Poëphagus] grunniens*), Richard Palmieri notes that religious attitudes, in this case Buddhist rather than Hindu, have helped to shape the role that the animal plays in the life of the peoples of the Himalaya.[10] Molly Debysingh has found that spatial patterns in poultry keeping and eating in India can only be explained with reference to cultural factors and that, furthermore, in areas of heavy population pressure where one would expect to find high levels of poultry keeping, cultural deterrents inhibit the use of poultry as a cheap food source.[11] Even with wildlife populations, ideologies come into play, as Gary Dunbar illustrates in his article on ahiṃsā and *śikar* (hunting).[12] A prime example of this is the banning of hunting for the year 1974–75 by Gujarat and other Indian states in commemoration of the 2,500th anniversary of the attaining of *nirvāṇa* by Mahāvīra, the Jain *tīrthaṅkara* (teacher) who preached the philosophy of ahiṃsā.

It is, however, a domesticated bovine that best illustrates the complex interface placed by history, environment, and culture between man and animal in the subcontinent. In few other regions of the world is the influence of one animal felt in so many different facets of life as is that of the cow in India. Cattle represent one of the major resources of India. With only 3 percent of the world's land area, India has approximately one-third of the world's cattle population. The estimated 180 million cattle[13] form the largest concentration of domesticated animals anywhere in the world and, as might be expected, are an important element in the Indian agricultural economy. The cow contributes to the 18.8 million tons of milk produced annually in India[14] and, though the buffalo accounts for 60 percent of this total, it is cow's milk and products made from cow's milk, such as *ghī* (clarified butter), that are favored by the Indian population. In addition, the cow is valued as a producer of male calves—by no means an inconsequential role in a country where the main source of traction has traditionally been the bullock. Even today, despite the advent of the tractor and the diesel pump, for the great majority of Indian farmers it is the bullock that pulls the cart and plough and that draws water from the well for irrigating the fields.

Both cow and bullock make other contributions to the life of the Indian peasant. Over the centuries, many parts of India have suffered severe deforestation. Wood is scarce and dung is used as a substitute fuel. A study by the National Council of Applied Economic Research indicates that some 25.5 percent of all energy consumed in rural India comes from dung, used mostly for cooking in the form of dung cakes.[15] Similarly, the use of dung for manuring the fields, an integral part of traditional farming practice in South Asia, is increasing in significance with rising costs of chemical fertilizers. The usefulness of the animals continues even after death since, despite much wastage, hides and bones are processed for leather, fertilizer, and meal, while beef is eaten by many people, the Hindu taboo notwithstanding.

Yet, beyond this purely utilitarian function, which is the lot of most domesticates, cattle play a much wider role in the life of India, the repercussions of which are far-reaching. To many Hindus, the cow is not just an animal to be exploited economically but is viewed with a reverence and respect normally reserved for the senior members of one's own family, or even for the lesser gods of the Hindu pantheon. Indeed, the extent to which reverence for the cow in particular and cattle in general is carried can be seen in Hindu religion and mythology. The bull in the form of Nandi, for example, is the sacred vehicle of Shiva, the Destroyer of the Hindu Trinity, and is found at the entrace to all Shiva temples throughout the land. The bull is also linked with other Hindu gods, such as Dyaus, Agni, Rudra, and Indra. Krishna is represented as the "cowherd" god, and many of the tales and legends of his life pursue this motif, linking him to the cow and pastoral settings. The cow, itself a symbol of fertility, figures prominently in Hindu myth and legend, appearing as Surabhi, the mother of all cows, as Kāmdhenu, the cow that grants all desires, or as the home of all the gods save one, Lakshmī, who finding no space remaining in the body of the cow is forced to take up her abode in its dung. The existence of *goloka*, the place of the cows, which ranks above the heavens of mere gods and mortals, the primeval ocean of milk which is churned by gods and demons, the frequent allusions to the cow in the sacred literature—all attest to the prominent position of the cow in the religious life of India.

It is not only in religion and mythology that we find these unique attitudes towards cattle expressed, for they are apparent in the daily rites, practices, and habits of the Hindu people. The cow is worshipped at various festivals. The bulls of Shiva are suffered to roam the streets unmolested. Cow's milk is used in temple ritual, while to *pañcagavya*, the "five products of the cow" (milk, curds, ghī, urine, and dung), are ascribed certain magical and medicinal properties. Cow dung is used in the home for the ritual purification of the hearth. Most high-caste Hindus will not countenance even the slaughtering of cows, let alone the eating of beef. As a result, the cow has been a traditional focus of communal discord between Hindu and Moslem ever since the arrival of Islam in India. More recently the cow-protection movement in India and the legislating of antislaughter laws in various states have brought the issue out of the realm of folk culture and religion and into the modern political arena, with demands from traditionalists for a total ban on slaughter throughout the country. In April 1979, for instance, Āchārya Vinoba Bhave, a much-respected Hindu reformer and spiritual heir to Mahātma Gandhi, went on a hunger strike to pressure the central government into imposing a nationwide ban on cow slaughter in India. At first Prime Minister Morarji Desai claimed to be helpless, arguing that antislaughter legislation lay within the realm of the states' powers and not that of the Delhi government, but such was the popular outcry in support of the Āchārya that Desai was forced to intervene. Only after Bhave was assured by the Prime Minister that the central government would push for a national ban on the slaughter of cattle with all possible speed did he break his five-day fast.[16] There are, of course, groups opposed to the cow-protection movement, including economists who see traditional attitudes toward cattle as a hindrance to India's economic growth, religious minorities such as the Moslems who have raised the question of their constitutional right to make a living, even though this be from butchering, and communists who find cow protection totally irreconcilable with their political philosophy.[17]

These attitudes, beliefs, and practices associated with cattle, often referred to as the "sacred cow concept" of India, form a significant and visible element on the Indian cultural scene. They raise, moreover, questions of broader interest to the student of culture and society concerning the nature of cultural evolution. How did the various elements of this particular complex of culture traits originate? Are they unique to the Indian subcontinent and Indian culture, or do they have parallels elsewhere in the world? If so, where do they occur and how can this distribution be explained? What effect have these elements had on the development of Indian culture? Can they contribute anything to our understanding of the nature of cultural growth and development?

As might be expected, considerable differences of opinion exist among scholars who have considered these issues. Theories of the origin of the sanctity of the cow, for example, are as varied as attitudes toward the cow found in India today. The traditional viewpoint is offered by W. Norman Brown, a Sanskritist of renown, who concludes that the sacred cow concept is the product of at least five elements, most of which are religious in nature. After a thorough examination of the Sanskrit literature, Brown cites as contributing factors the role of the cow in Vedic ritual; the figurative use of words for cow in the Vedic literature and their subsequent literal interpretation; Vedic prohibitions against violations of Brahmans' cows; the ahiṃsā concept; and the association of the cow with the Mother Goddess cult.[18] Brown's conclusions as to the religious nature of the doctrine of the sacred cow appear to have widespread support among Sanskritists and other students of Indian culture.[19]

Other theories of origin, however, invoke a variety of political, economic, and religious factors. Hindu reverence for the cow and related practices such as adherence to the ahiṃsā concept and the avoidance of beef are, for example, seen by some as conscious attempts on the part of Brahmanical Hinduism to counter the dominance of Buddhism throughout India during the early centuries of the Christian era.[20] Alternatively, the crystallization of Hindu dogma concerning the sanctity and inviolability of the cow has been interpreted as a response to the Moslem invasions of the eleventh and twelfth centuries A.D. and ensuing efforts to establish Islam in India. More recently, it has been suggested that the origins of the sacred cow concept lie in the political and economic policies deliberately pursued by Indian states in the past. The appearance of the concept in Indian history coincides with the large-scale sedentarization of agriculture and the rise of urban states such as Magadha. As the argument goes, prohibitions on cow slaughter and beef eating were imposed on the mass of the Indian peasantry by the politically dominant urban elites of these newly emergent states in order to create surpluses for their own benefit.[21]

The major challenge to traditional views of the sacred cow concept, however, is posed by anthropologist Marvin Harris who, in questioning religious interpretations of the doctrine in two articles published in 1965 and 1966, initiated a stimulating and ongoing debate concerning the nature of the sacred cow concept, its origins, and its role in Indian society.[22] Harris, adopting the position of cultural ecology first developed by Julian Steward,[23] suggests that the sacred cow doctrine and its manifestations in Indian society result from "positive-functioned" and "adaptive" processes stemming from the Indian ecological setting rather than from the negative influences of Hindu religious thought. The entire cattle complex, he argues, forms part of a naturally selected ecosystem and can be understood in terms of the "techno-environmental base, Indian property relations and political organization," rather than in relation to religious concepts such as ahiṃsā. Harris maintains that taboos against beef eating and the slaughtering of cattle are not aberrant patterns of behavior rooted in irrational ideologies but are traits totally consistent with an ecological balance. The entire aura of sanctity which surrounds the cattle complex is "thoroughly circumscribed by the material conditions under which both man and beast must earn their livings."[24]

Thus for Harris and his supporters, the sacred cow is a myth. The aura of sanctity surrounding the cow in India is not so much a product of Hindu religious philosophy but is rather a cultural mechanism to protect a valuable economic resource—a mechanism that itself reinforces and lends rational support to apparently irrational practices, such as adherence to ahiṃsā and rejection of beef. From this point of view the cultural values associated with the so-called sacred cow of India are not detrimental to the Indian economy. To the contrary, they are advantageous in that they act to maintain a delicately balanced functional system under the techno-environmental conditions pertaining to the Indian subcontinent.

But the validity of Harris's views on the sacred cow, indeed, of the entire methodological structure of cultural ecology as used by many anthropologists, is open to serious question. Paul Diener and Eugene Robkin, for instance, are highly critical of the functional-ecological approach. They take cultural ecologists to task for ignoring, or at best being unaware of, the value of the evolutionary approach in understanding cultural systems:

> The ecologists' failure to recognize the importance of an autonomous, complementary evolutionism in anthropology is faulty on several grounds. First, although claiming inspiration from biological theory, ecological anthropologists fail to appreciate the crucial distinction between functional-ecology and evolution so widely accepted in the life sciences. Second, the attempt to account for origins by reference to ecological data often results in superficial research. Third, functional-ecological arguments are logically suspect. Fourth, ecological speculations about origins turn us away from the real historical record; as a result many problems open to explanation in terms of developmental history are instead ecologically "explained away." Finally, though ecologists have sometimes been critical of the motives and social values underlying other research methodologies, the social forces which have resulted in the ecological "explosion" in American anthropology have gone largely unexamined.[25]

While most of Diener's and Robkin's remarks are directed toward the use that has been made of the functional-ecological approach in general, and towards Harris's explanation of the Islamic prohibition on pork in particular, their criticisms are equally valid for ecological interpretations of the sacred cow of India. In a later study, Diener, Robkin, and Nonini focus on the particular problem of India's sacred cattle, joining Heston, Dandekar, Bennett, and others in rejecting Harris's functional explanation of the Indian cattle complex.[26]

Simoons, for instance, in his discussion of specific issues arising from the sacred cow controversy, finds Harris's account of both the origins of the sacred cow concept and its functioning in Indian society to be seriously flawed. Indeed, in his earlier writings on the subject Harris ignores completely the question of origins, but in later works he proposes a scenario in keeping with his functional-ecological views. Cattle came to be viewed as sacred, Harris argues, at a time of widespread environmental degradation and intense ecological pressures on human and cattle populations alike. Under such conditions, so great was the importance of cattle for peasant survival that taboos against slaughter and beef eating developed as "the cumulative result of the individual decisions of millions and millions of individual farmers."[27] Simoons, however, sees major inaccuracies in this hypothesis, not the least of which is its misrepresentation of the historical record. Much of the destruction of the environment in northern and western India following 1,000 B.C., he contends, was man induced and was largely the result of overgrazing. It did not make sense for farmers to ban beef eating and to stop slaughtering cattle in order to "remove temptation," as Harris claims, thus increasing cattle numbers and further intensifying pressures on the environment. Simoons cites literary evidence for cow sanctity to show that the concept was imposed from above and not developed independently by farmers.[28]

Furthermore, Simoons continues, Harris's ecological approach does not adequately explain the facts of the situation relating to cattle in India today, since his commitment to a techno-environmental determinism leads him to ignore or dismiss evidence contradictory to his stated position. Harris sees ahiṃsā, for example, as a positive-functioned trait deriving "power and sustenance from the material rewards it confers on both men and animals."[29] Yet the ahiṃsā concept does operate in situations where no possible material rewards are forthcoming, and does not function along the purely mechanistic lines that Harris envisages. As Bennett notes, Harris fails to view religion as a "strategy of action" and does not become involved in the religio-political question, even though religion and other ideologies are important influences on Indian attitudes and behavior toward cattle.[30] Rather than seeing the ahiṃsā concept and the rest of the cattle complex as positive-functioned and adaptive, Simoons holds that "the proper framework in which to place the sacred-cow controversy is one which permits traits to be positive-functioned, negative-functioned, or both and which allows for human choice among alternative cattle policies and systems."[31]

However one chooses to view the cattle complex of India, inherent in the continuing controversy that surrounds the sacred cow concept are questions of wider significance, one of which concerns the relationship between Hindu religious values and economic development in India. The impact of value systems on economic achievement is, of course, of interest to a wide range of social scientists, and it is to a sociologist that one must turn for a definitive statement of the problem, both in general terms and in relation to India. In his classic study of Protestantism, Max Weber raises the question of why, despite the continued existence of elements of capitalism in both India and China in the past, economic development in these eastern civilizations did not follow the course it took in western Europe.[32] After a detailed analysis of the Indian situation, he comes to the conclusion that the answer is to be found in the "essentially negative" effects of Hinduism and the "completely traditional and anti-rational" nature of the caste system.[33]

This view that the values associated with Hinduism and Hindu society exert a negative influence on progress in general and on Indian economic development in particular is expressed in numerous writings reaching as far back as the eighteenth century. In the late 1790s, for example, the French missionary and scholar, Abbé Dubois, wrote:

> It is, to my mind, a vain hope to suppose that we can really very much improve the condition of the Hindus, or raise their circumstances of life to the level prevailing in Europe. . . . as long as it is in the nature of the Hindus to cling to their civil and religious institutions, to their old customs and habits, they must remain what they have always been, for these are so many insurmountable obstacles in the path of progress. . . .[34]

More than one hundred years later, the same feeling is stated much more succinctly by the economic historian, Vera Anstey, who, in her study of the economic development of India, writes, "The religious tenets and practices of Hinduism and Muhommadanism have strictly limited economic development in the past, and influence fundamentally future potentialities."[35] In a more recent attempt to discern and assess the impact of Hinduism on India's economic growth, Vikas Mishra concludes that although the institutions and attitudes of Hinduism aided economic growth and welfare in the past, after the beginning of the impact of western industrialization they acted as brakes on economic growth.[36]

Similarly, according to K. William Kapp, an economist who accepts the influence of cultural values on economic behavior:

> It can hardly be doubted that Hindu culture and Hindu social organization are determining factors in India's slow rate of development. It is not only the lack of capital resources or skilled manpower which impedes the process of economic growth but non-secular and pre-technological institutions and values such as the hierarchically organized caste system, the limited or static levels of aspirations, moral aloofness, casteism and factionalism—to name only a few of the major barriers.[37]

The impact of some of these nonsecular and pretechnological institutions and values are well illustrated in Kusum Nair's vignettes of the human factor in economic development in India.[38]

Not all authors, however, accept this view that cultural values and institutions associated with the Hindu religion have acted to hinder economic growth and development. Many observers, commenting on the distinction in India between real life and the theoretical model of society set out in the corpus of ancient literature, have noted that the Indian peasant is an eminently practical individual who has little difficulty in reconciling his spiritual beliefs with his material well-being. The alleged contrasts between the "spiritualistic" East and the "materialistic" West, argues Milton Singer, though no doubt valid, may well have been overemphasized. The pursuit of wealth and material power represents a distinct motif within the Hindu tradition, exemplified by the worship of Lakshmī, the goddess of wealth, and the existence of *artha* (wealth) as one of the four roads to salvation. Thus, cultural values may not be as great an obstacle to economic development in India as had previously been thought.[39] This position is supported by the economic historian M. D. Morris, who comments that, despite finding his conclusions disturbing, "there is no precise definition of a 'Hindu value system' that can be identified as a significant obstacle to economic growth or change."[40]

As Morris indicates, the apparent confrontation between values and economic growth in India is an artificial one, especially since both concepts can be, and often are, defined in essentially subjective terms. Schwabe points out, for instance, that outsiders viewing India's cattle complex through Western-colored glasses see irrational cattle practices as "the unique product of peculiar, anachronistic and 'undesirable' aspects of the Hindu religion."[41] Western society, on the other hand, prides itself on being organized by a strict material rationality, with practical and utilitarian considerations underpinning all aspects of social and economic behavior. But what of "the land of the sacred dog," as Marshall Sahlins calls it? His discussion of American food taboos and the American clothing system clearly illustrates that the so-called rationality of Western society is as much rooted in cultural bias as the "irrationality" of Hindu attitudes toward the cow.[42] With this in mind, it becomes apparent that any set of beliefs and values forms an integral part of a cultural whole and cannot be set up in opposition to economic or social behavior or isolated from its total cultural context. Sahlins proposes an alternative to viewing cultural design in terms of component systems such as economy, society, and ideology, each with its own unique characteristics and objectives, and the whole arranged in a hierarchy determined by views of functional dominance and necessity. He suggests, instead, that we should assume a perspective which reflects an awareness of "the diversity of cultural emphases . . . made more precise by the understanding that these represent differing institutional integrations of the symbolic scheme. Here the economy appears dominant . . . there everything seems 'bathed in a celestial light' of religious conceptions."[43]

In India, it seems that the celestial light of religious conceptions casts an aura over land, people, and economy alike. Religious beliefs influence the manner in which Indian society perceives and utilizes its environment and resources; they play a critical role in the selection of one cultural system, however functional or dysfunctional, over alternatives. They are also important in the adaptation and survival of existing systems. Thus the question at hand is not how traditional values affect economic behavior or how irrational beliefs inhibit economic growth but rather what is the role of belief systems in social and economic change.[44] What impact have Hindu beliefs and related behavior patterns had on cultural evolution in India? How have they responded and adapted to the vagaries of Indian political, social, and economic history? What are the contradictions and conflicts produced by this particular set of ideologies or strategies of action in Indian society, and how have they been resolved, if at all?

In the light of these questions and the more specific issues raised in the sacred cow controversy, the animal homes of India become more than just anachronisms, moribund relics of dying cultural values. In reality, as this writer has fully come to appreciate, goshalas and pinjrapoles are living manifestations of attitudes and patterns of behavior that lie at the very heart of Indian society, even though they are survivals of a cultural tradition that appears to extend back through the centuries to the earliest days of Hindu civilization. Their continued existence today, at a time of rapid economic and social change, makes them of even greater interest because, as an element of the Indian cultural landscape, they act as mirrors of the society that conceived and fostered them. Not only do they embody a traditional set of beliefs and values, but changes in their nature and functions may be expected to reflect more general processes at work in Indian society. They become, to some degree, a signpost to the future as well as a record of the past.

Combining traditional methods of cultural-historical analysis with field research, this study examines the nature, origins, and evolution of the animal home in India. A preliminary discussion of terminology, current ideas concerning origins, and variations in the types of animal homes existing in India today is followed by a consideration of contemporary distributions. The historical development of the institutions is traced from earliest times up to the present, and the discussion of modern animal homes is continued in detailed case studies of selected goshalas and pinjrapoles. On the basis of this, the organization and functions of animal homes and their role in economic planning and development are examined. The work concludes by placing animal homes in the broader perspective of the sacred cow concept, of resource utilization, and of the role of religious values in influencing economic and social behavior.

This study adds to our knowledge of a little-known feature in the cultural landscape of India and, in doing so, should help to unravel one of the more obscure aspects of the Indian cattle complex. In addition, it sheds some light on the sacred cow concept itself and contributes to a better understanding of the intricate and tenuous bonds linking man, culture, and animals in the Indian subcontinent.

II

Goshalas and Pinjrapoles—Forms in Modern India

A herd of cows, with horns painted silver, gallops through the narrow streets of a bazaar while vegetable vendors keep a wary eye on their stalls. An old, wrinkled grandmother painfully lowers herself to her knees and presses her forehead to the dusty ground in the wake of the vanishing herd. A seventy-year-old man walks half a mile in the heat of a plains summer to massage the limbs of crippled cows that are too weak to stand. Two magnificent zebu bulls, the trident of Shiva and the circle of Sūrya emblazoned on their flanks, lock horns in a primitive test of strength. A farmer, his very livelihood threatened by the failure of the monsoon rains, brings his starving bullocks to a place of refuge where they will be watered and fed.

These cameos of Indian life belong not to the world of Marco Polo or Ralph Fitch but to the India of the present. They are all incidents related in some way to goshalas and pinjrapoles, institutions that are very much part of the contemporary cultural scene. According to HArbans Singh in a survey undertaken for the Central Council of Gosamvardhana, there were in 1955 an estimated 3,000 such institutions in India maintaining some 600,000 head of cattle,[1] to which could be added a variety of animals ranging from deer and dogs to camels and cats.

Translated literally, *goshala* means "place for cows," thus giving some indication of the institution's traditional function in providing shelter for and maintaining old and disabled cattle. The word *pinjrapole*, on the other hand, appears to be derived from the Sanskrit *pañjara* and *pāla*, meaning "cage" and "protector," respectively,[2] and the prime role of this institution is to serve as an asylum or refuge for animals of any species or, where medical treatment is provided, as an animal hospital.

In India today, the words *goshala* and *pinjrapole* are often used interchangeably or even together to describe the same institution, and as a result some confusion exists both as to their nature and their functions. Despite this popular usage, on the basis of observations in the field I suggest that the traditional distinctions between the two institutions are still valid. Whereas the goshala is concerned with cattle only, the true pinjrapole acts as a refuge for all animals, and it is here that one finds both wild and domestic creatures. Further confusion arises from the fact that the Hindi word *goshala* has several meanings; it can be taken literally to designate any place, such as a field or pasture, where cattle are to be found; it can mean a dairy; or it can be used in the more formal sense of an institution set up for the preservation, protection, and development of cattle. Unless otherwise stated, it is this last interpretation that is used throughout this study.

Some institutions, such as the Calcutta Pinjrapole and the Nasik Pinjrapole, go under the name of pinjrapole but in reality function as goshalas. In these cases, the institutions were initially founded as pinjrapoles offering shelter to all animals but subsequently came to devote their attention exclusively to cattle, though retaining their original names. Alternatively, where goshalas have developed dairy herds, the distinction is often made between the milking cows and the pinjrapole section containing the sick, disabled and useless animals—hence the use of the term *pinjrapole goshala*. In both instances, these institutions may be regarded as goshalas in that they now serve only cattle.

Although the words *goshala* and *pinjrapole* are by far the most commonly used throughout India for animal homes, there do exist other terms and phrases to describe them. They may be known, especially in the south, as *gorakshan* (*gosamrakshan*) *shāla* (home for cow protection) or simply called *gorakshan sangham, samiti,* or *sabha* (cow-protection societies). Occasionally the more poetic *gomandir* (cow temple), *gosevā mandir* (temple for service to the cow), or *gosevāśram* (*āśram* for service to the cow) may be found, while the government has added to the list in recent years by creating the *gosadan*, a collection center for useless and decrepit cattle. *Paśu anāthālaya* (animal orphanage) or *jīvdayā apāng sansthā* are terms that may be used instead of pinjrapole.

Despite this variety in terminology, the main purpose of the animal home in India, according to the literature, would appear to be cow protection. Satish Das Gupta, for example, traces goshalas to the gorakshan samitis (cow-protection societies) of village India. These first came into existence, he argues, when Indian culture was based mainly on village communities. Rather than slaughter cows that were unremunerative through old age or disability, groups of villagers joined together to look after the animals. Funds were provided by charity, and farmers and breeders contributed by paying taxes on their produce.[3] Sardar Datar Singh, noting that most goshalas and pinjrapoles came into existence during the last two centuries, writes that they were founded to:

> . . . stop any further deterioration of the cow as a giver of milk and draught power and to give necessary protection to the cow from indiscriminate slaughter and to house those infirm animals which have served throughout their lives, during their old age.[4]

Shukla, on the other hand, feels that the institutions stem from man's "spontaneous desire to protect them (cattle) from sufferings in old age."[5] Y. M. Parnerkar, Honorary Chief Editor of *Gosamvardhana*, puts forward another point of view, writing in an editorial:

> The institutions of Goshalas and Pinjrapoles were started primarily to look after old and infirm cattle. The need stemmed from the fact that under imperial subjugation it was hardly expected of the alien government to take interest in the protection and preservation of the cattle wealth of the country. It therefore devolved on the philanthropists to do something for such cattle and out of this moral obligation felt, came into being the Goshalas and Pinjrapoles.[6]

Gandhi views pinjrapoles as "an answer to our instinct for mercy"[7] and goshalas as "a refuge for all worn out and maimed cattle."[8]

Some of these ideas are supported by the findings of H. J. Makhijani, who, in a survey of goshalas and pinjrapoles carried out for the Central Council of Gosamvardhana, reports that reasons given for the establishment of animal homes include the provision of care and shelter for disabled, old, and maimed animals (Nasik Panchayati Pinjrapole, Nasik, Maharashtra, founded 1879); the protection of old and disabled cows, bulls, and bullocks during the 1891 famine (Śri Goshala, Bhagalpur, Bihar, founded 1891); looking after uncared-for cattle let loose by their owners (Jharia-Dhanbad Goshala, Jharia, Bihar, founded 1920); providing training facilities in cattle management (Gandhi Vidhya Mandir Goshala, Sardar Shahr, Rajasthan, founded 1950); and providing milk for the patients in the Nature Cure Ashram (Ashram Goshala, Uruli Kanchan, Maharashtra, founded 1948).[9]

The limited size of Makhijani's survey, however, tends to reduce its usefulness since not only does it fail to include true pinjrapoles (today the Nasik Pinjrapole functions as a goshala) but it makes no distinction between various types of goshalas—a distinction of no mean consequence in terms of the role that they play in contemporary India and of their origins and historical development. An analysis of field data that I gathered recently from some 100 goshalas and pinjrapoles shows that at least six different kinds of animal homes are to be found in India today, namely: (1) the pinjrapole; (2) the temple goshala; (3) the court goshala; (4) the *vania* goshala; (5) the Gandhian goshala; and (6) the gosadan. Although the specific nature of their activities may vary, in all of these institutions the protection, preservation, and development of cattle is an important, if not the ultimate, objective. Even in pinjrapoles, with their emphasis on animal life in general, cattle are usually the most numerous among the animal population. Yet each of these institutions represents a discrete element on the broader canvas of cow protection in India, often reflecting different motivations, cultural traditions, or even political and economic pressures. A closer look at these six forms will illustrate the variety of animal homes that exist in India today.

The Pinjrapole

Inscribed in some prominent position over a door or gateway in all pinjrapoles can be found the words *ahiṃsā parāmo dharma* (ahiṃsā is the greatest of religions). In this aphorism is summed up the entire raison d'être of pinjrapoles, for it is the extension of ahiṃsā and the related concept of *jīv-dayā* (compassion for life) to embrace all animal life that accounts for the presence of the institutions in India today. A typical pinjrapole thus offers asylum to a wide range of creatures. In May 1975, for example, the animal population of the Chhapriali Pinjrapole in Bhavnagar District, Gujarat, consisted of 6 milk cows, 752 old, lame, and dry cows, 233 bullocks no longer capable of working, 14 working bullocks, 356 calves under one year old, 6 useless female buffalo, 72 male buffalo calves, 44 goats, 105 kids, 402 sheep, 9 horses, 1 hare, 1 monkey, and 1 deer. If animals seen in other pinjrapoles are added, this list may be expanded to include rabbits, chickens, and peacocks (Rajkot Mahajan Pinjrapole), cats, dogs, and pigeons (Ahmedabad Pinjrapole) and camels and nilgai[10] (Sayla Pinjrapole).

Despite the more unusual creatures found in pinjrapoles, the main efforts of the institutions today are directed toward domestic herd animals, goats, sheep, and especially cattle. The Ahmedabad Pinjrapole, for instance, reported an intake of over 10,000 cattle and 12,000 sheep during 1973. Some of these animals find their way into pinjrapoles because their owners are unable to keep them, not because their productive lives are over. Salvageable animals are often reclaimed and placed in milk herds or returned to farmers, thus adding economic dimensions to the animal-protection aspect of pinjrapoles' activities. But these are minimal in number compared to the majority of herd animals found in the institutions, animals that are sick, injured, or too old to be of any economic value.

Of special interest, however, are those animals with no possible economic uses for man, since their presence in pinjrapoles illustrates the true spirit and purpose of animal homes in India. In the heart of Old Delhi, for instance, opposite the Red Fort and close to the bustle of Chandni Chowk, is a pinjrapole devoted entirely to the welfare of birds. Founded in 1929 as an expression of the Jain[11] community's concern for ahiṃsā, the Jain Charity Hospital for Birds' sole function is to treat sick and injured birds brought there from all over the city. Many Jain families have actually set up centers in their own homes in various parts of Delhi, to which sick and injured birds in need of treatment are taken and then sent on to the hospital by messenger.

The hospital, located inside the premises of a Digambara Jain temple and supported entirely by public donations administered through the temple committee, receives some thirty to thirty-five birds daily. Most of these are pigeons with wounds or fractures incurred in the city's heavy traffic, although diseases ranging from blindness to cancer are treated by the hospital's resident veterinarian. All birds, both wild and domestic, are accepted for treatment by the hospital with the exception of predators, which are refused on the grounds that they harm other creatures and thus violate the ahiṃsā principle. Incoming birds are treated in the dispensary on the second floor of the hospital (the first contains the staff quarters and the grain store) and are placed in one of the numerous cages with which this level is lined. As birds improve they are taken to the third floor, where they convalesce in a large enclosure having access to the open sky. A special cage is provided on this floor for the weak, maimed, and paralyzed to separate them from the other birds. When birds die in the hospital, they are taken in procession to the nearby Jumna and are ceremoniously placed in the waters of that sacred river.

The Jain Bird Hospital is unique and certainly not typical of the pinjrapole in India, yet all such institutions harbor birds and many, especially in Gujarat, have *pārabaḍīs* (fig. 1). The pārabaḍī is a large, ornate birdhouse, often reaching seven meters or more in height, usually made of wood, and elaborately carved and painted. It takes the form of a roofed, houselike structure or platform placed above ground level on a supporting pillar, so that it is out of the reach of cats and dogs. On this platform, grain is regularly set out to feed the birds, usually pigeons, which flock around any Gujarati village or town. Pārabaḍīs are found in streets and bazaars throughout Gujarat, as well as in pinjrapoles.

A similar structure is seen in the *kabūtriya*, the birdhouse that often occupies the central location in village and town squares in central

Fig. 1. *Birdhouse* (Pārabaḍī) *at Nadiad.*

Gujarat (fig. 2). Although it is usually much larger than the pārabaḍī, the kabūtriya fulfills the same function (as its name implies, *kabūtar* = pigeon), serving as a roost for pigeons and other birds. These impressive and ornate buildings are constructed with funds donated by the local community, especially the business castes, in the name of jīv-dayā and *dharma* (religion). Devout individuals will regularly leave grain and water in the kabūtriya for consumption by birds. This type of birdhouse is not usually found in pinjrapoles, although in some institutions smaller and less elaborate kabūtriyas may replace the pārabaḍī.

The provision of food for animals in pinjrapoles is not restricted to birds alone. Some institutions in Gujarat have a *kutta-kī-rotī* (bread for dogs) fund, from which a daily meal of *rotī*, the flat, unleavened bread of India, is provided for the ubiquitous pariah dogs. This treatment differs markedly from the usual harsh lot of the canine elsewhere in the subcontinent.

Of all the features of the pinjrapole, however, perhaps the most unusual is the *jīvat khān* or insect room. European travelers to India had commented upon their presence in pinjrapoles as early as 1583,[12] and a complete description is provided by Burnes who, writing of his visit to the pinjrapole in Surat in 1823, says:

> By far the most remarkable object in this singular establishment is a house on the left hand of entering, about twenty-five feet long, with a boarded floor, elevated about eight feet; between this and the ground is a depository where the deluded Banias throw in quantities of grain which gives life to and feeds a host of vermin, as dense as the sands on the sea-shore, and consisting of all the various genera usually found in the abodes of squalid misery.
>
> The entrance to this loft is from the outside, by a stair; which I ascended. There are several holes cut in different parts of the floor through which the grain is thrown. I examined a handful of it which had lost all the appearance of grain; it was a moving mass, and some of the pampered creatures which fed on it were crawling about on the floor—a circumstance which hastened my retreat from the house in which the rest of the vermin is deposited. The 'Pinjra Pol' is in the very midst of houses in one of the most populous cities in Asia: and must be a prolific source of nightly *comfort* to the citizens who reside in the neighbourhood; to say nothing of the strayed few who manage to make their way into the more distant domains of the inhabitants.[13]

This most remarkable object no longer exists at Surat, though the visitor to the present pinjrapole there may see a round, well-like structure sunk into the ground where, until recently, grain was deposited for insects and other vermin.

In Ahmedabad, however, on a narrow street in the old quarter of the city, the jīvat khān still survives (fig. 3). Here, in a building especially constructed for the purpose by the Ahmedabad Pinjrapole Society, three

Fig. 2. *Birdhouse* (Kabūtriya) *at Mandal.*

Fig. 3. *Insect Rooms* (Jīvat Khāns) *at Ahmedabad.*

rooms are set aside for insects. Devout Jains from all over the city bring the dirt and dust and sweepings from their houses and leave them outside the door of the insect room. For the religious-minded, this dirt contains insect life and to discard it might lead to the destruction of this life, thus violating ahiṃsā. Daily this debris and waste is placed inside the insect room by pinjrapole workers. It is sometimes supplemented with grain donated by the faithful for the sustenance of the room's inhabitants. When the jīvat khān is full, the door is locked and the room is kept closed for periods of up to fifteen years. By this time, it is felt, all life in the room will have come to a natural end and its contents can be disposed of without possible harm to its former inmates. The room is opened and the residual of dust, decaying grain and, presumably, organic matter is, with the business acumen typical of the vania castes, sold as fertilizer.

While not all pinjrapoles maintain insect rooms today, they do exist in Nadiad (Kaira District) and Palanpur (Mehsana District) as well as in Ahmedabad. In Palanpur, the pinjrapole even employs a man to make daily rounds of the town's streets crying out "jīv-dayā," at which Jaina housewives bring him dirt and sweepings to be deposited in the jīvat khān. It is in such places that the original character of pinjrapoles has been preserved. Despite the increasing importance of cattle and other herd animals in their populations, they continue to afford shelter to the animal world at large, and it is this which sets the true pinjrapole apart from the other types of animal homes existing in India.

The Temple Goshala

Of the various types of goshalas that can be seen in India today, perhaps the most ancient are those attached to Hindu temples. Not only does the cow figure prominently in Hindu mythology and folk tradition, but cow worship (*gopūjā*) is an integral part of Hindu ritual. At many temples, cows may be kept on the premises so that devotees can perform pūjā or obtain *darśan*[14] of the cows. In addition, cow's milk and milk products have long played an important role in Hindu ritual and ceremony, a feature that can be traced back to the earliest of the Vedas and which apparently antedates the sanctity of the cow itself. Milk is used to bathe the image of the god, be it the *liṅga* (phallic emblem) of Shiva or the *mūrti* (image) of Krishna. Ghī is burned in the lamps that light the inner sanctum of the temple, the lamps used for *ārtī* by the Vaishnavas,[15] and the lamps lit at *Dīwālī*, the Festival of Lights, in honor of Lakshmī and Rāma. Milk is the main ingredient in *pañcaṃrit*, the "nectar" made from the products of the cow which is distributed to participants in temple ceremonies. Milk and milk products are also used to make various *prasād*, offerings of sweets daily presented in temples to the gods by their worshippers.

The basic role of the temple goshala, therefore, would appear to be the provision of milk and cows for ceremonial use in Hindu temple ritual. Not all temples maintain goshalas and where they do, productive animals often outnumber useless ones. Some of the larger temples, such as the Jagannath Temple at Puri in Orissa, however, report upwards of 2,000 head of cattle, many of which are nonproductive, in their goshalas.

The Court Goshala

An institution that may well rival the temple goshala in its antiquity is the court goshala, attached to the households of the colorful *rājas* and princes who, until Independence, ruled over the myriad of independent kingdoms and states that dotted the face of India. Even in the heyday of the British Rāj, more than five hundred states were governed by independent rulers, both Hindu and Moslem, under the aegis of British paramountcy. Many of the Hindu princes maintained goshalas for both religious and economic purposes. In Banaras, for example, that most orthodox of orthodox states, it was the custom for the Mahārāja to be awakened by the lowing of a cow, which was brought to his bedroom and prodded for that purpose. During a visit by the Mahārāja to the neighboring state of Rampur, this custom posed his host some problem since the guest's bedroom was on a second floor. The Nawāb of Rampur supposedly solved his dilemma by buying a crane and hoisting a cow up to his guest's window—much to the dismay of the palace's sleeping inhabitants and, needless to say, of the cow as well.

With the passing of the Rāj, these remnants of India's pomp and splendor died out too. The court goshala has largely gone the way of the princely states. The present Mahārāja of Banaras does still maintain such an institution, but this relic of the past is a mere shadow of the times when the size of the court goshala was a measure not only of the wealth but also of the piety of the true Hindu prince.

The Vania Goshala

The third type of goshala that exists in India is the vania goshala. The word *vania* means "merchant" or "shopkeeper," but it is also used throughout India to refer to the business and trading castes in whose hands commercial activities are concentrated. The vania community has been instrumental in setting up and managing many of the goshalas found outside the western states of India, so the term has been adopted to refer to institutions organized by the business classes.

As is true of all goshalas, the wide range of animals characteristic of pinjrapoles is absent from the vania institution, for only cattle are accepted. Even the buffalo, which is widely recognized as being a better milk animal than the cow, is barred from the vania goshala. In fact, when the question of accepting buffalo was raised with goshala members, some became quite vehement, stating emphatically that because the animal is unclean and is undeserving of the same respect as the cow, it would never be admitted into their institution.

The objective of the vania goshala, according to those who support the institution, is gorakshan (cow protection and development) and gosevā (service to the cow) stemming from considerations of dharma (religion). This, so the argument goes, is laid down in the scriptures and is the proper duty of every righteous Hindu. *Gai hamārī mātā hai!* (The Cow is our Mother!) She must, therefore, be protected from the neglect of the ignorant and from the depredations of the Moslems and, formerly, the British imperialists, who think nothing of cow slaughter and beef eating. The cow is a useful animal that should be treated well during the last years of her life. One does not, an Agarwal vania from Calcutta explained, kill one's mother once her useful life is over and she can no longer contribute anything to the household. Even so with Mother Cow—she gives us life through her milk and is a tender and faithful companion. She deserves better than death once her useful economic life is over, and so she should be cared for and fed in her old age. The cow, moreover, is a holy animal beloved of the gods; to kill her not only violates ahiṃsā but is a sin worse than the killing of a man.

That these are popular sentiments among the trading castes of India is reflected in numbers, for the vania goshala is the most common type of animal home existing in the country today. Its numerical dominance, combined with its extensive distribution in the subcontinent, has meant that the vania goshala has received much attention in recent years at the expense of the other types of institution.

The Gandhian Goshala

What is called here the Gandhian goshala is a recent revival of an apparently ancient institution. Many regard Mohandas Karamchand Gandhi, the Mahātma, as one of the giants of the twentieth century, and it is true that no other individual has had such a sweeping impact on the course of Indian history during this time. Undoubtedly he is remembered more for being the father of the Indian nation who led his country to freedom from Imperial Britain than for his impact on Indian philosophy. Yet for Gandhi, political freedom without spiritual freedom was meaningless, and spiritual freedom was to be sought in the teachings of the Hindu sages of old. Central to his philosophy was the desire for a return to the traditional values and practices of ancient India, and cow protection was important to this theme:

> The central fact of Hinduism is cow protection. Cow protection to me is one of the most wonderful phenomenon [sic] in human evolution. It takes the human being beyond his species. The cow to me means the entire subhuman world. Man through the cow is enjoined to realize his identity with all that lives. . . . Protection of the cow means the protection of the whole dumb creation of God. . . . Cow protection is the gift of Hinduism to the world. And Hinduism will live as long as there are Hindus to protect the cow.
>
> Hindus will be judged not by their *tilaks,* not by the correct chanting of *mantras,* not by their pilgrimages, not by their most punctilious observance of caste rules but by their ability to protect the cow.[16]

Thus cow protection and service to the cow was basic to Gandhi's philosophy. Just as Gandhi himself adopted the simple life of the ascetic and founded several ashrams, centers where devout Hindus could retreat to a life of meditation and contemplation, so a part of this life was the goshala. Sabarmati Ashram at Ahmedabad, Uruli Kanchan near Poona, and Sevagram southwest of Nagpur were all founded by Gandhi and all had goshalas,[17] which fulfilled several functions. They were living symbols of Hindu reverence and respect for the cow, they maintained old and useless cattle (although they did not actively seek them out from the public in the manner of the vania institutions), and they provided fresh, unadulterated milk for the inmates of the ashrams.

Today there are two types of Gandhian goshalas in India. Those attached to ashrams, which are often religious centers set up by *gūrūs* or *sunyāsīs* (religious teachers), tend to be fairly small and serve the same ends as in Gandhi's original ashrams. Near Gauhati in Assam, for example, located on the slopes of a hill crowned by the famous Kamakhya Temple and overlooking the Brahmaputra River, is the Anand Niketan Ashram. Founded by Śrima Sadanand Mayi Giri, a female devotee of Shiva who continues to serve as spiritual head of the institution, the ashram has attached to it a small goshala with seven milk cows, one breeding bull, eleven calves, and three useless cows. Milk from the goshala is used by "Sanghamātā" (Mother of the Society) and her devotees, while any surplus is sold on the local market.

The second type of Gandhian institution is that run by traditional educational establishments. Here, however, while still fulfilling the same functions, it often serves the additional purpose of providing training to students in modern dairying methods and techniques. A *gūrūkul* is a traditional Hindu school, and the Gurukul Goshala at Chitorgarh in Rajasthan provides a typical example of this type of institution. Attached to an ashram, the gūrūkul was founded by one Swami Vratanand, a follower of the Arya Samaj, which is an extremely conservative, right-wing orthodox Hindu religious movement that has, understandably, assumed political overtones. The current school principal is also a *swāmī* (religious teacher) and an Arya Samajist. The school curriculum, which is taught in Hindi, includes Sanskrit and studies of the sacred literature as well as the basics of reading and writing. The Gurukul Goshala was founded primarily to provide milk for the school but also to help instill a sense of respect for the cow into the students of the school. Although the goshala does not normally accept useless cattle from the public, it does maintain eight useless cows along with a milk herd of thirty-six animals. It also reports that in 1969, when Rajasthan was experiencing drought, herders from Marwar migrating with their cattle to the Indore region of Madhya Pradesh left in the goshala several animals too weak to proceed, rather than have them perish en route.

Even though gūrūkuls are traditional institutions, they sometimes follow advanced practices. The Virani Goshala at Shadagram near Mangrol in Saurashtra, although attached to a traditional-style school, is surprisingly modern in outlook, even to the point of playing recorded music to its milk animals. In some of the larger institutions, especially in colleges and universities, this process of modernization has proceeded to such an extent that former goshalas would now have to be considered as modern dairies. This has happened, for example, in Dayanand Vivekananda College in Ajmer, Rajasthan, and in Banaras Hindu University, both originally Hindu educational institutions now offering a wide range of degree courses to all comers. Yet even in Banaras (Varanasi), with the original goshala now a dairy farm operating under the auspices of the Department of Animal Husbandry, cows that become too old to lactate are not killed or sold—at present eight out of a herd of seventy-five are nonproductive but are maintained at an annual cost of some Rs. 1,200.

The Gandhian goshala thus differs markedly from the other animal homes under discussion. Like the temple goshala, it is attached to a parent institution, but it provides milk for local consumption rather than for ritual purposes. It has the same aims as the vania goshala, seeking to preserve ancient Hindu values concerning the cow, and it supports useless cattle, though it does not accept such animals from the general public. Given the relatively small number of Gandhian goshalas found in India today, however, this type of institution is of considerably less significance than the pinjrapole, the temple goshala, and the vania goshala.

The Gosadan

The most recent in origin of all the animal homes in India, the gosadan is a direct result of the Indian government's attempt to reconcile religion and economics, to solve the dilemma posed by a constitution that specifically affords protection to the cow in a land where useless animals apparently compete with a starving population to eke out a meager existence. As part of the first Five Year Plan, covering the period 1951–1956, the government decided to create reserves (gosadans) in isolated areas where nonproductive cattle could be sent, thus locating them where they could do little harm to fields and crops. Under the Gosadan Scheme, all old, infirm, or otherwise useless cattle were to be collected by officials at the village level and sent to segregation camps deep in the jungle and forest areas. At the gosadan there were to be facilities for housing the cattle, which would be grazed on the surrounding lands. At each camp a small tannery was to be built for flaying the dead animals and for processing the carcasses for bonemeal and tallow. Thus the gosadan would work in several ways; first, it would remove stray and useless cattle from areas where they could damage crops, sterilize them so that they could not breed, and isolate them from productive cattle to reduce the spread of disease; second, it would concentrate them in one locale so that maximum use could be made of their dung and carcasses; and third, it would provide an alternative to slaughter.

Several such gosadan centers have been set up in various states in India. The success or failure of this plan is considered elsewhere, but at this point it should be noted that the gosadan differs from all the other animal homes discussed previously in that it is a state-sponsored and funded institution created as part of an overall economic planning policy to meet the specific problem of surplus cattle.

It is apparent that the animal homes of India are of greater consequence and complexity than has been suggested by Makhijani's survey, or than envisaged by scholars who have viewed the institutions as marginal to the problem of the cow in India. Quite to the contrary, it would seem that here the religious and the economic, the sacred and the profane, merge in institutions seated at the nexus of behavior patterns that hold great import for Indian society. Ahiṃsā, dharma, temple ritual, Gandhi's philosophy, the realities of modern economics—all are linked in some way with the goshalas and pinjrapoles of India and thus call for further analysis. First, however, the animal homes themselves require closer attention. The preceding discussion has outlined the presence of six distinct types of animal homes in India, each of which fulfills differing needs and each of which responds to a differing set of factors. In that these factors are subject to spatial variations, an examination of the distributions of animal homes in India will enable us to identify and evaluate them more effectively.

III

Distributions, Locations, and Spatial Hierarchies

Distributions

The animal home is essentially a feature of northern and western India. The distribution map of animal homes (map 1) shows that they lie in a belt extending northeast from Saurashtra through Rajasthan to Haryana and the Punjab, then swing southeast into Uttar Pradesh and Bihar. Southward into the Deccan, the occurrence of animal homes declines rapidly, with Maharashtra and Madhya Pradesh characterized by significantly lower densities. A few scattered institutions are found in the southern and eastern states of the peninsula, but extensive areas are totally devoid of them. Similarly, animal homes are noticeably absent in the mountains to the north of the plains belt. Jammu and Kashmir and the tribal hill states in the east have no goshalas and pinjrapoles, while the vast sweep of the Himalayan foothills is virgin territory broken only by the presence of a handful of institutions in the hill districts of Uttar Pradesh and at Kathmandu and Darjeeling. In the northeast, the few goshalas are, with the single exception of the Shillong institution, concentrated on the lowlands of the Brahmaputra Valley.

Within the main belt of goshalas and pinjrapoles, three regions stand out as cluster areas of high density, the most important of these being in Gujarat. Here, in the Kathiawar peninsula (which comprised the former state of Saurashtra) and northeast Gujarat, it is not uncommon to find districts with as many as thirty or more institutions, the highest concentrations found anywhere in the subcontinent. A second focus can be discerned in the north, consisting of the northeastern districts of

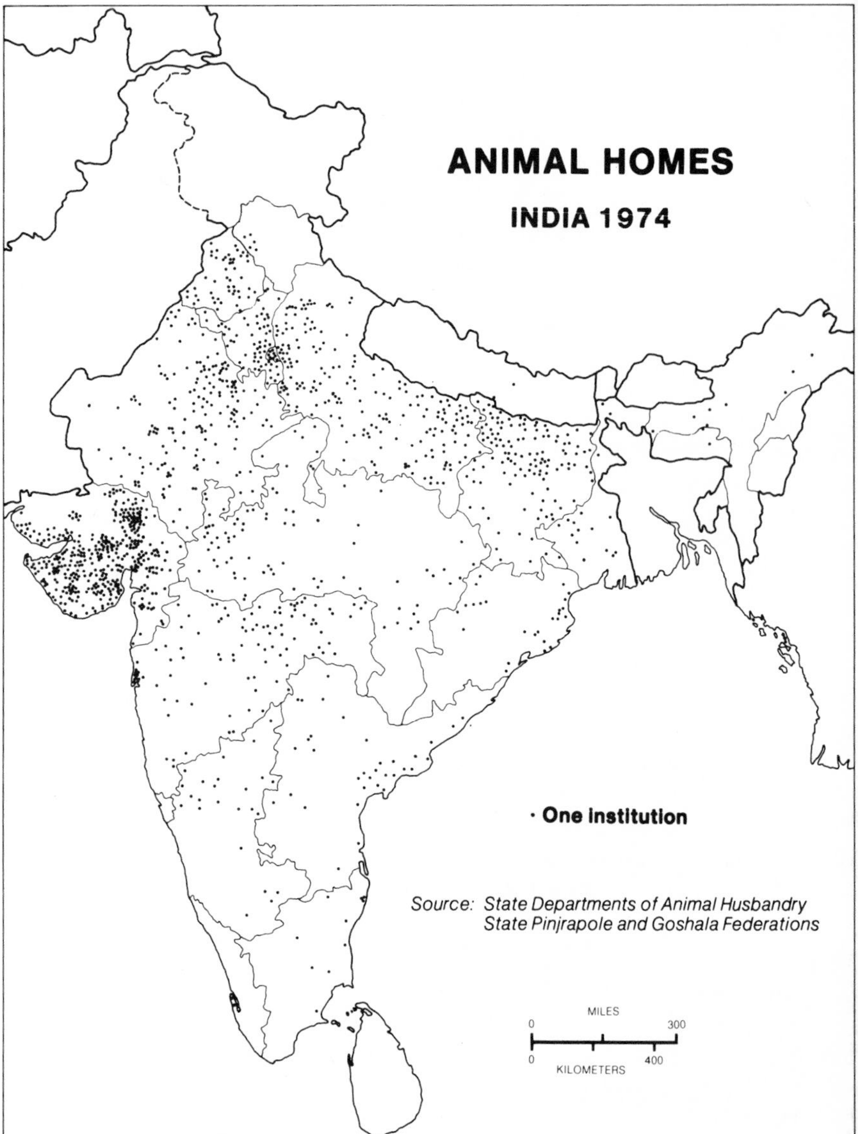

Map 1. *Animal Homes in India, 1974*

Rajasthan, Haryana, the Punjab, and some western districts of Uttar Pradesh. The third cluster area is found on the *terai* and alluvial plains of the lower Ganges valley, encompassing the districts of eastern Uttar Pradesh north of the Goghra River, the northern districts of Bihar, and a belt extending along the Ganges itself in Bihar.

Elsewhere, patterns become less distinct. Only Bombay stands out as a major center outside the main belt. Minor concentrations are found in western Madhya Pradesh around Ujjain and in Maharashtra along the upper reaches of the Tapti River, but these can in no way compare with the northern and western foci. Two lesser clusters are seen along the east coast, one on the deltaic lowlands of the Mahanadi River and the other along the mouths of the Godaveri and Krishna Rivers. Eastern Madhya Pradesh, the interior uplands of Orissa and Andhra Pradesh, Tamil Nadu, Kerala, Karnataka, and coastal Maharashtra south of Bombay are all virtually devoid of institutions.

How, then, can the basic characteristics of this distribution pattern be explained? The main concentrations of animal refuges coincide with the alluvial lowlands of the Indo-Gangetic plains, with regional variations from this pattern being attributable to physical features such as the low-lying and seasonally inundated mud flats of the Rann of Kutch or the desert lands of the Thar. Except for a few scattered institutions in the foothills, the Himalayas form a sharply delineated northern boundary, while in the peninsula many of the areas characterized by a marked absence of animal homes coincide with the uplands of the Eastern and Western Ghats. Yet, despite this predominantly lowland distribution, the negative influence of relief alone cannot account for the actual presence and concentrations of goshalas and pinjrapoles throughout the subcontinent.

Environmental factors do influence the functioning of animal homes in India. One reason frequently put forward for their existence is that they serve to protect the country's vast cattle resources at times of natural calamities, suggesting that adverse physical conditions, usually climatic in nature, might have been responsible for their emergence. Several goshalas and pinjrapoles in famine-hit districts of Gujarat reported that their intake of livestock, especially cattle, during the 1972–1974 drought, when the monsoons failed for three years in succession, rose by as much as 100 percent over normal years. The high densities of animal refuges in Gujarat, therefore, might well reflect an environmental setting where a low and highly variable precipitation makes for periods of unreliable agriculture and scarcity of fodder supplies. Similarly, in the summer of 1974, areas of Bihar south of the Ganges were stricken with drought, while most of the state to the north lay submerged under the floodwaters of tributaries descending from the Himalayas, an unenviable situation that played havoc with agricultural activities. Yet if Gujarat and Bihar, both areas with high concentrations of animal homes, frequently experience drought or flooding, the same cannot be said of the third cluster area, the Punjab. Here, in one of the richest agricultural states in the Union, moderate annual precipitation combined with an extensive system of irrigation effectively diminishes the impact and incidence of such climatic disasters. Natural calamities might, therefore, intensify the cattle-protection role of goshalas and pinjrapoles in various parts of the country, but they cannot be invoked as a causal factor explaining either their existence or their distribution in India.

If environmental determinants are unsatisfactory in accounting for the spatial distribution of goshalas and pinjrapoles, cultural factors offer more promise. The basic pattern calls to mind the traditional division between Aryan and Dravidian India. Their absence from the Deccan and concentrations on the northern plains suggests that the institutions derive from an Aryan (Vedic) tradition rather than from the older Dravidian cultures of the south. This view is supported by the apparent failure of the animal home to penetrate the mountain belt in the north or to gain acceptance among the tribal peoples of the eastern littoral, who were never effectively brought into the mainstream of Hindu culture.

Such generalizations, however, are of limited value. While defining the basic outlines of the distribution of animal homes in India, they mask regional variations that reflect significant cultural associations. The differing forms and functions of animal homes, moreover, make them subject to a variety of locational factors, each of which contributes to the composite distribution pattern apparent in map 1, and each of which warrants further consideration.

Locational Factors

Of all the types of animal homes under discussion, the pinjrapole shows the most distinct spatial pattern in its distribution over the subcontinent (map 2). The greatest concentration of pinjrapoles is found in Gujarat—more precisely in Kutch, in the Kathiawar peninsula, and in the northeastern districts of the state. Outside of Gujarat, minor concentrations are seen in neighboring areas of Maharashtra and in the northeastern districts of Rajasthan, while only a handful of institutions are found elsewhere in the country, located mainly in major urban centers such as Calcutta, Madras, and Bangalore.

The map of pinjrapoles in India also shows important differences in the distributions of pinjrapoles and pinjrapole goshalas. It is in Gujarat

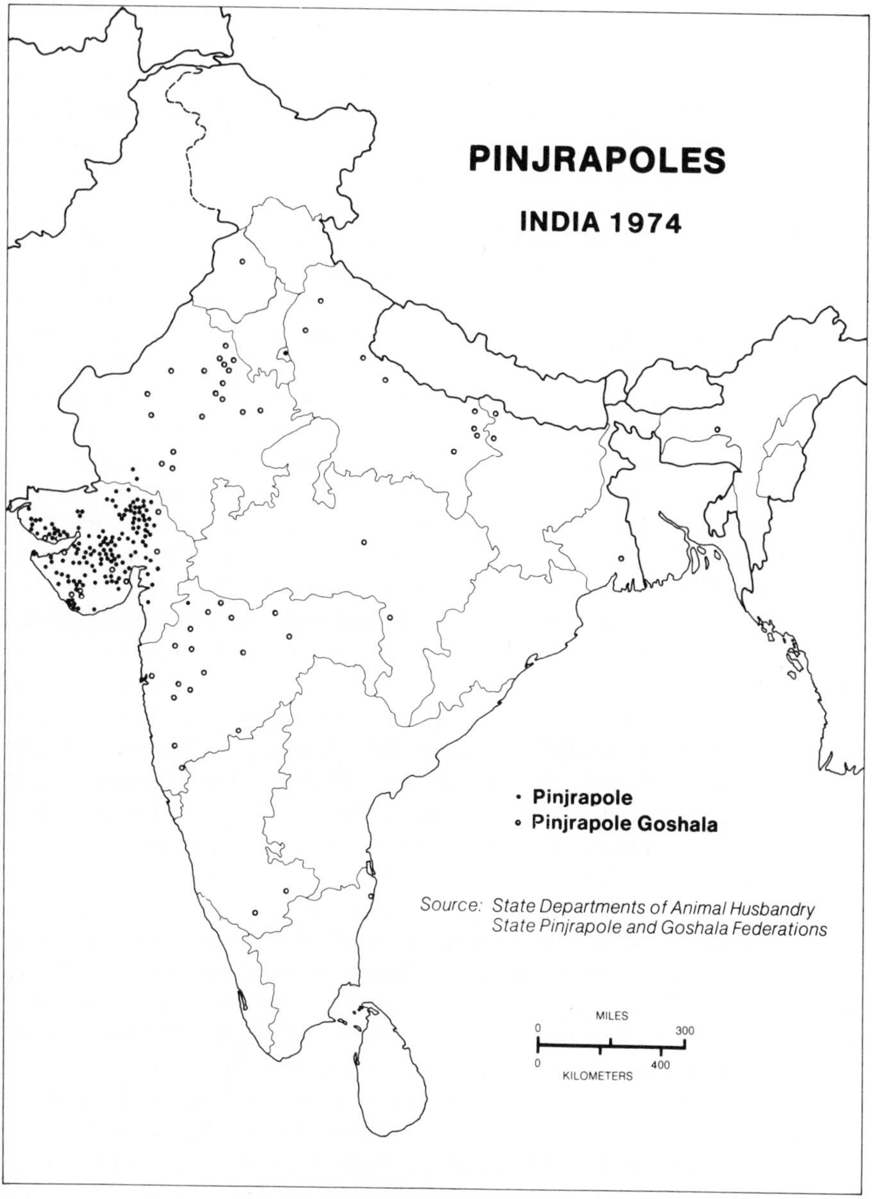

Map 2. *Pinjrapoles in India, 1974*

that the pinjrapole appears to have maintained its original character, for some 97 percent of the institutions which today offer asylum to all species of animals lie within that state. The majority of pinjrapoles outside this core area function as goshalas, directing their efforts toward cattle. Yet many of these pinjrapole goshalas were established as true pinjrapoles and, in their origins, reveal ties with Gujarat. The Nasik Pinjrapole, for example, was founded by the Gujarati merchant community of the city, while in Ahmednagar, the pinjrapole was established at the turn of the century by Gujaratis and Marwaris. Some of these institutions outside of Gujarat are even required by their rules and regulations to publish their annual reports in the Gujarati language.

If origins and distributions suggest that pinjrapoles are essentially Gujarati institutions, a closer look at Gujarat itself reveals even more specific cultural associations. There exists, for example, a strong correlation between pinjrapoles and the distribution of the Jain population within the state. Districts such as Kutch, Surendranagar, and Ahmedabad show not only the highest percentages of Jains in their population but also the highest concentrations of pinjrapoles.[1] This pattern holds true elsewhere in the country where the institutions are found, for both Rajasthan and Maharashtra are regions of strong Jain influence (map 3).

An analysis of distributions, therefore, indicates that pinjrapoles are Gujarati institutions closely associated with Jainism. The overwhelming concentration of the institutions in Gujarat, the survival of their traditional animal-protection functions in the state, and the erosion of these functions in the rest of the country all point to Gujarat as the area central to the origins of the pinjrapole. Here have developed those values that conceived and nurtured the institution; from here those values, and perhaps the pinjrapole itself, have been carried across the subcontinent by the Gujarati communities that today engage in business and commerce in cities and towns throughout India.

Vania goshalas, on the other hand, lack the distinct regional pattern shown by pinjrapoles, conforming more to the basic distribution of animal shelters in India (map 4). The three areas of highest density stand out clearly, but institutions are found from the Punjab to Tamil Nadu, from Gujarat to Assam. It is, moreover, in areas peripheral to the main distributions that an explanation of this pattern is to be sought. In Assam, for example, six of the nine goshalas functioning in the state in 1974 were vania institutions, and of these six, five had been established by Marwaris[2] within the last hundred years. In neighboring Meghalaya, the sole goshala was founded in Shillong in 1948 by the Marwari Panchayat, which consisted of ten Agarwals and one Jain. The early Marwari community in Shillong itself originated in the Shakwati region

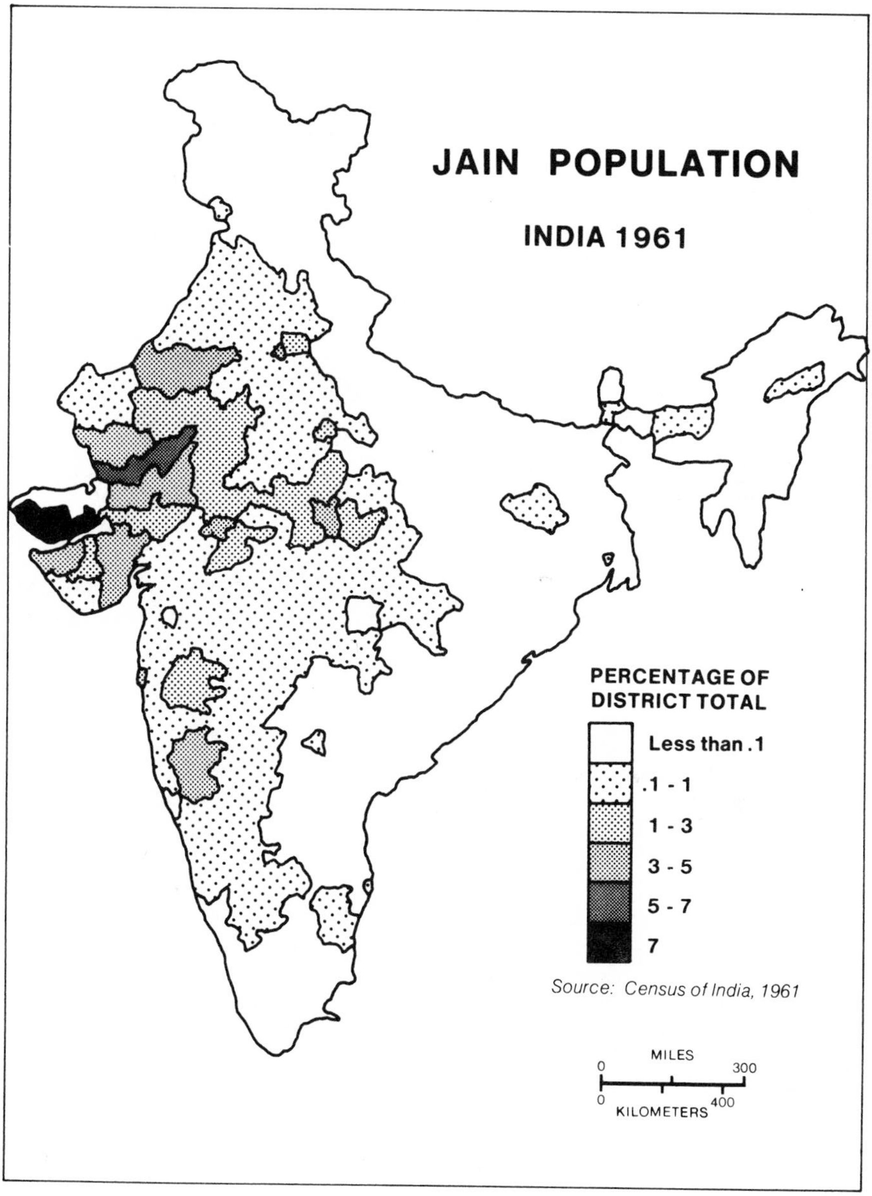

Map 3. *Jain Population, 1961*

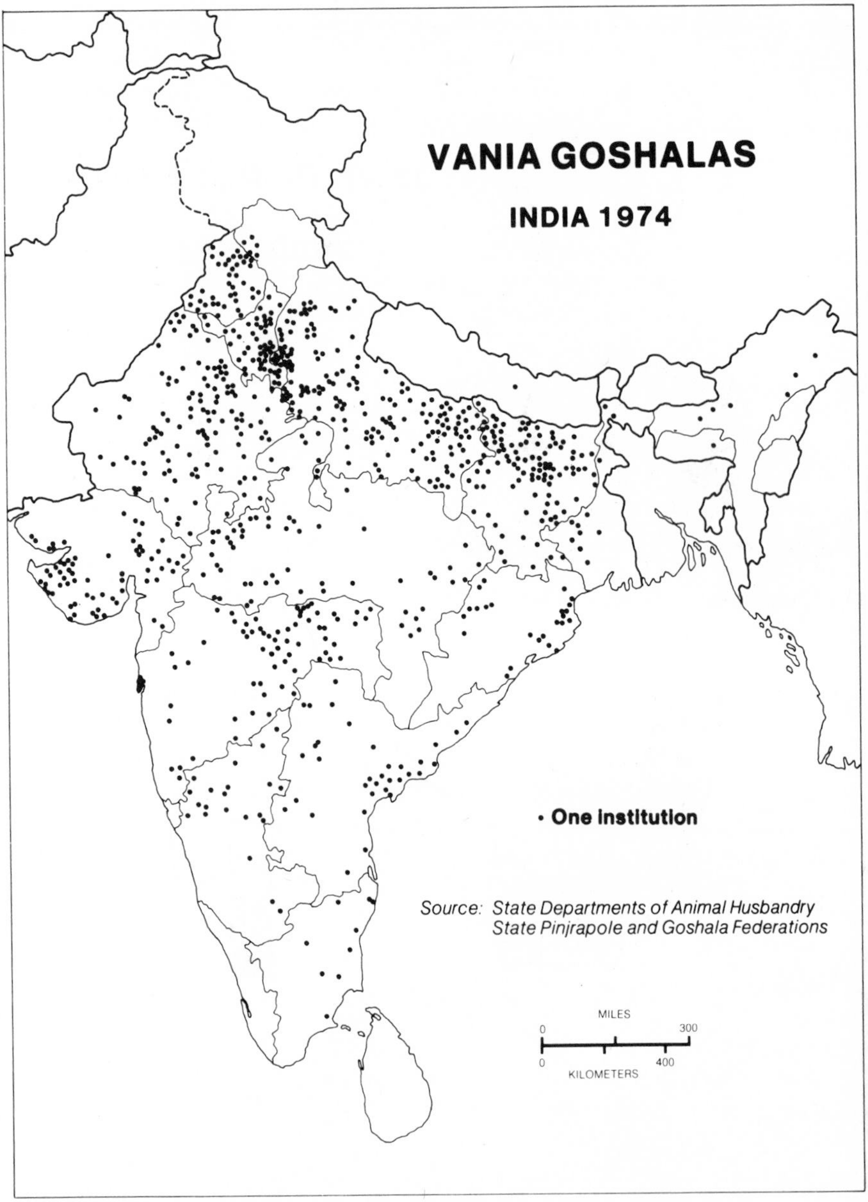

Map 4. *Vania Goshalas in India, 1974*

of Rajasthan, and it migrated to what was then Assam around the 1870s. Similarly, the Darjeeling-Siliguri Goshala in West Bengal dates to 1898, when it was set up by local Marwaris whose historical roots lie in the Jaipur, Jodhpur, and Bikaner regions of distant Rajasthan. Even in Nepal, the few goshalas found in this mountain kingdom are Marwari in origin.

The distribution of vania goshalas in India, therefore, would appear to be closely linked to the presence of Marwari vania communities, for the associations apparent in Assam and Nepal hold true for the rest of the country. It is significant that Rajasthan, the homeland of the Marwaris, lies within one of the three areas of high densities of animal homes, and that Marwaris form an important element in the urban population of a second in Bihar. The evidence suggests that late nineteenth-century migrations of Marwaris were responsible for the introduction of vania goshalas into the more remote areas of the north and east, and it seems likely that similar migrations, though at different times, can account for the presence of the institutions in other parts of the country as well.

Temple goshalas are, of course, attached to existing religious institutions, but the factors influencing the location of temples are so varied that only the most general observations, usually of a historical nature, are possible. Areas of strong Vaishnava influence such as western India, where Krishna reputedly ruled at Dwarka and where Vallabhāchārya carried out his mission in the early sixteenth century, show a high incidence of both Krishna temples and temple goshalas. The locations of some of these, the Śri Nathji Temple at Nathdwara and the Ranchorji Temple at Dakor, for example, may commemorate incidents, real or imaginary, taken from the legend and lore of Krishna.[3] Beyond this, however, it becomes difficult to view the distribution of temple goshalas in anything less than individual terms. In Gujarat alone, for instance, there exist goshalas attached to temples dedicated to Krishna, Shiva, Sūrya, Hanumān, and various forms of the Mother Goddess (map 5).[4]

In the same manner, the distribution of Gandhian institutions is related to the locations of ashrams and educational establishments. These reflect decision-making processes that are individualistic in nature and are in no way linked to the intrinsic functions of animal homes. Why Gandhi should found ashrams at Sabarmati and Wardha, and why the Shadagram School should be built in Saurashtra after being forced to relocate from West Pakistan at the time of Partition are matters of history, personal preference, and chance rather than of systematic locational processes.[5]

Of all the animal homes in India, the only type to be planned and thereby subject to a set of specific locational factors is the gosadan. Ideally these are situated in isolated rural areas far enough away from

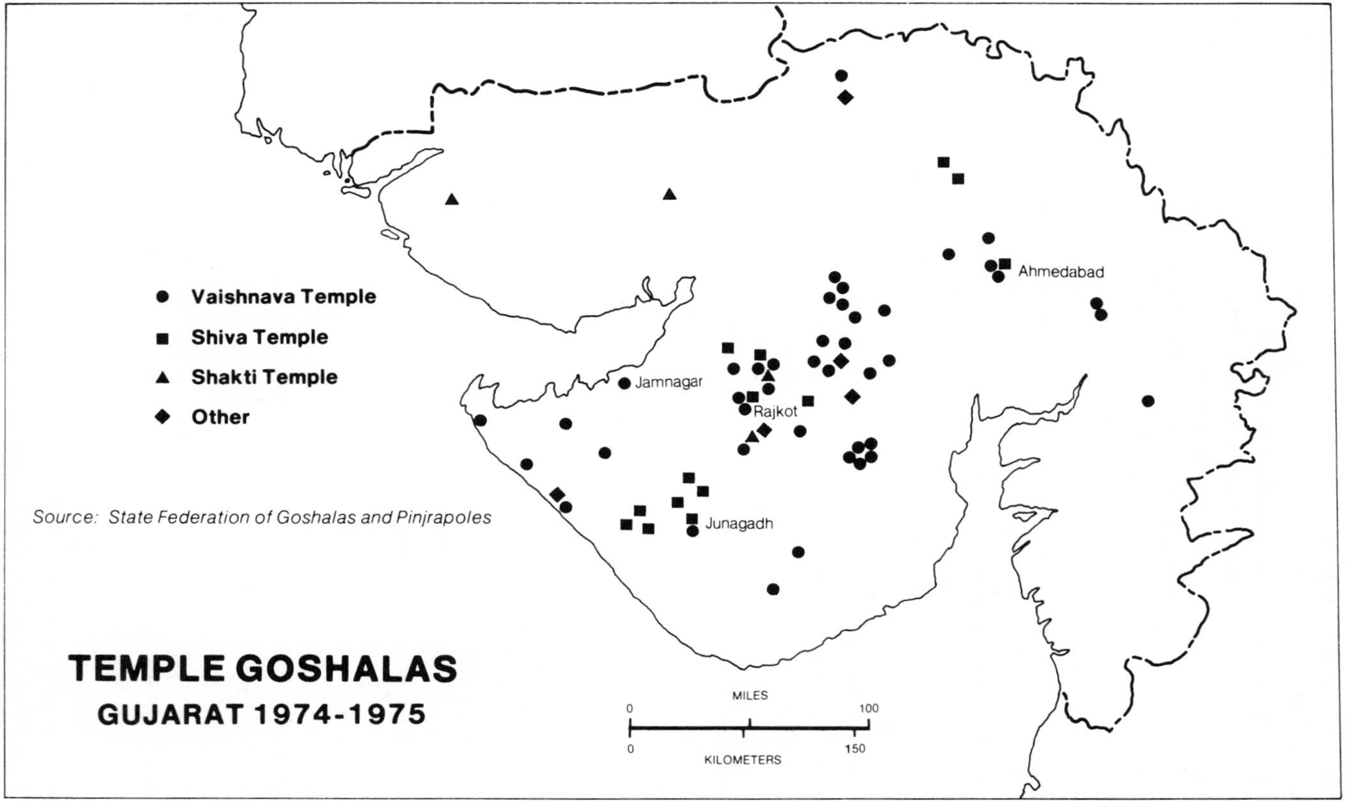

Map 5. *Temple Goshalas in Gujarat, 1974–1975*

settlements so as not to interfere with agricultural activities, but close enough and accessible enough to limit transportation costs, since the cattle populations are shipped in by road. Adequate pasture is also required to reduce the need to bring in fodder supplies. Difficulties in obtaining suitable locations have, in combination with other factors, detracted from the successful operation of these institutions.

Spatial Hierarchies

The animal homes of India are essentially urban institutions. Some temple goshalas, Gandhian institutions, and the gosadans do not fit this pattern, but pinjrapoles and vania goshalas are invariably located in cities and towns, since their main source of public support is the local urban vania community. In times of need, however, the support potential in smaller towns is greatly enlarged by the practice of approaching former residents who have moved to larger cities, such as Ahmedabad and Bombay, for aid. Smaller goshalas and pinjrapoles may also be helped by the larger and better-endowed institutions in the area which accept overflow cattle and sometimes even provide financial assistance.

Despite urban locations, most animal homes have extended their operations into neighboring rural areas. Many institutions, through donations and purchases, have acquired agricultural land. While maintaining a head office at their original premises, they have moved their herds to their rural branches. This has often been necessitated by lack of space, especially in the older institutions. Thus many animal homes with rural locations are in fact extensions of operations based in the cities.

It is possible to identify a hierarchy of urban institutions reflecting location, size, and what might be called spheres of influence—areas from which financial support is received. At the base of the pyramid are goshalas and pinjrapoles that are purely local in character. They depend entirely on local revenues and donations, while their animal-protection activities extend over a limited area in their immediate vicinity. Nonproductive animals usually outnumber the productive, with cattle populations of considerably less than a hundred. Economic activities such as milk production are poorly developed, and property and landholdings are minimal. Typically the institution is found in smaller towns, although outside the main concentrations of animal homes they may be situated in larger urban areas. These local institutions form the most numerous group in India, but many of the smaller homes, with limited funding and public support, are the most likely to succumb to adverse economic conditions and to cease functioning.

Above these are regional institutions with a broader resource base and a service area that may extend over several districts. They are usually located in major regional towns and, through various mechanisms, receive financial support from areas as much as one hundred fifty kilometers distant. Their assets include considerable land and property, which generate an assured income and result in much greater financial security. Their scale of operations is larger than that of the local institutions, and they are often characterized by well-developed economic functions, having sizable milk herds and even, in certain instances, making draught and milk cattle available for agricultural purposes. The Rajkot Pinjrapole in Gujarat and the Dibrugarh Goshala in Assam, both discussed in later chapters, are typical of this type of institution.

Finally, there are some four or five major institutions located in the premier cities of India, such as Bombay and Calcutta, whose size and scale of operations are of a different order of magnitude than those of the local and regional animal shelters. Their annual income exceeds a million rupees, while their assets include not only extensive holdings of land and property but often stocks and shares as well. Financial support is received from an area that may extend well beyond the state, as does the service area of the institutions, and all maintain four or five branches in neighboring rural districts. Economic activities are extremely well developed, with milk production and even cattle breeding often superseding the traditional animal-protection role of the institutions. Despite their size and prominence, however, in origin, organization, and functions they conform to the basic pattern characteristic of animal homes throughout India.

The large urban institutions are not the only animal homes that draw support from beyond their immediate region, for there exist a select few whose influence extends almost nationwide. These are the goshalas and pinjrapoles found at major religious pilgrimage sites. The cluster of goshalas around Mathura and Vrindavan in the heart of Krishna country, for example, are located there for the very reason that these locales are associated with the birth and early childhood of Krishna, one of the most important of the Hindu gods.[6] They are significant centers of pilgrimage for Vaishnavas, and indeed for all Hindus, and are visited every year by thousands of devotees who partake in the massive circulation of pilgrims throughout the land. Many of these people visit the goshalas for darśan of the cows and to offer donations, thus extending the area from which financial support is received to include most of the country. Similarly, pilgrims from all over India who visit the Jain temple city of Palitana in Saurashtra include, as part of their religious observances, donations to the nearby Chhapriali Pinjrapole (map 6). There exists, therefore, a

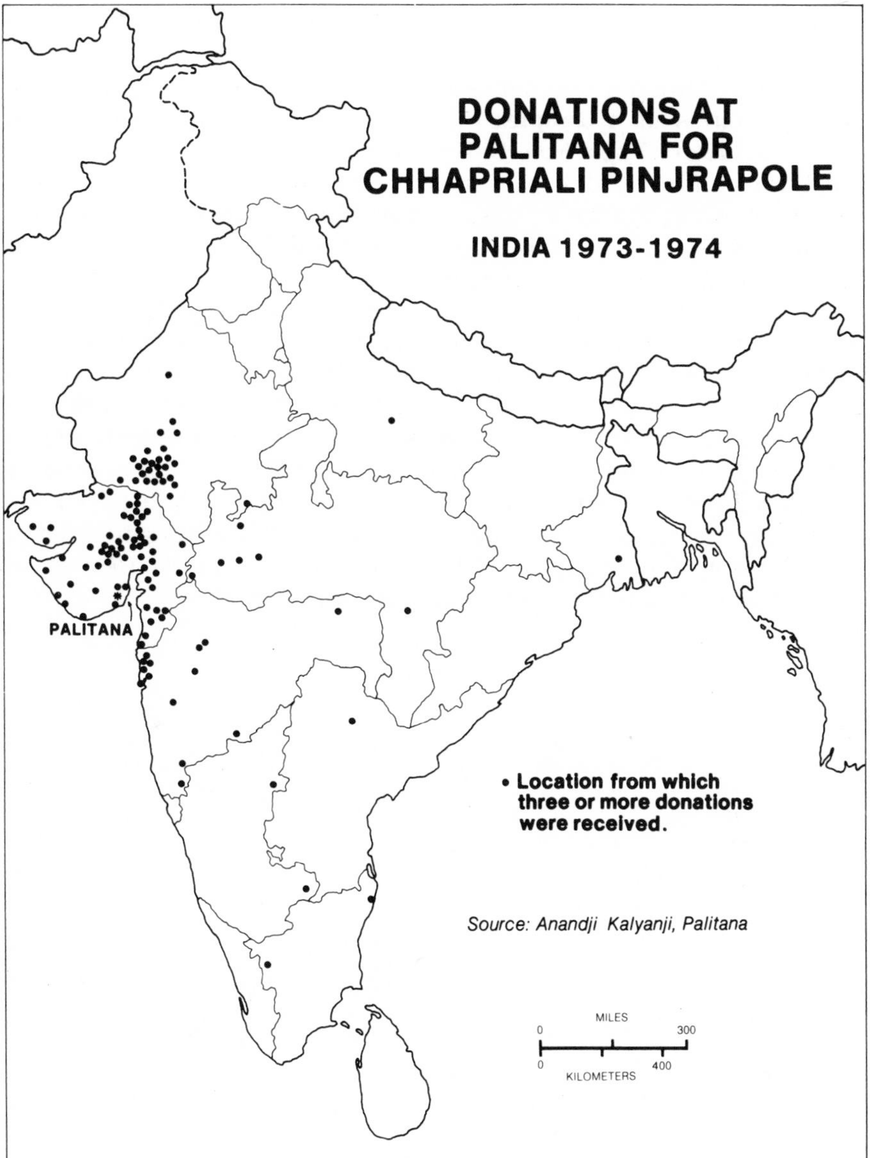

Map 6. *Donations at Palitana for the Chhapriali Pinjrapole, 1973–1974*

hierarchy of goshalas and pinjrapoles associated with religious sites, consisting of local temple institutions, those connected with shrines of regional standing, and a few that are linked to pilgrimage centers of national importance.

The distribution of animal homes in India can thus only be understood in terms of a complex of social, cultural, environmental, and historical factors. The northerly distribution and general accordance with the Aryan culture realm suggest that the animal homes are indeed Aryan in origin, while persistent associations with Jainism and traditional Hindu beliefs, with religious locations, and with specific northern vania communities, support the contention that the origins of the institutions are to be sought in the cultural traditions of northern India. To ignore origins is to ignore the forces that conceive and nurture, and the goshalas and pinjrapoles of today are but successors to institutions that may well have been in existence for over two millennia. A true understanding of the nature of animal homes is to be found not only in twentieth-century India but also in two thousand years of culture history. It is against the sweeping panorama of man and cattle in the Old World, the emergence of the sacred cow concept in India, the heterodox reactions against Vedic Brahmanism, and imperial conquests that the origins and historical development of goshalas and pinjrapoles must be viewed.

IV

Goshalas and Pinjrapoles—Origins and Historical Development

Early Cattle Cults in the Old World

The elevation of cattle to a position of religious or quasi-religious significance is restricted neither to India nor to Indian religious philosophies. The Indian cattle complex lies within a much more extensive area of the Old World in which a variety of traits such as cattle worship, cattle sacrifice, and cattle sports are found. Religious associations with cattle go back several thousand years, and Eduard Hahn suggests that these very associations provided the initial stimulus for the domestication of cattle by man.[1]

Whatever the validity of this view, available evidence points to the eastern Mediterranean as an early center of cattle cults. The oldest known remains of domesticated cattle occur in Greek Thessaly and are dated to ca. 8,500 B.P.,[2] but it is southern Turkey that provides one of the first instances of cattle depicted in a specifically religious context. James Mellaart's excavations at Çatal Hüyük on the Anatolian plateau provide ample evidence of thriving cult activities based on the bull, possibly as early as ca. 6,500 B.C.[3] From here and perhaps other sites in the area, bull cults may well have diffused throughout the eastern Mediterranean lands to achieve their greatest development in the later civilizations of Mesopotamia, the Indus Valley, Egypt, and Minoan Crete.

A detailed study of the origins and nature of early cattle cults lies beyond the scope of this study, yet there exist interesting parallels between the ancient Near East and India worthy of mention. Reverence for the bull as a symbol of masculinity and power, its position as a cult animal, and its apparent role in fertility cults are all features of the eastern Mediterranean that appear in some form or other in the Indus Valley civilization. Similarly, the representation of various sky and storm gods in Mesopotamia, Egypt, and the Levant as a bull, often as consort to a Mother Goddess, has its counterpart in the association of the bull with Dyaus, Rudra, and Indra in Vedic India.[4]

The cow, too, emerges as the symbol of a female deity. In pre-Dynastic Egypt, the Mother Goddess, regarded by some as one of the most important deities of the time, is identified with the cow on Naqada remains, one of her standards being cow horns mounted on sticks.[5] Hathor (fig. 4), originally worshipped in the form of a cow as a fertility symbol, came to be linked with the Sky-goddess, Nut, assuming the body of the celestial cow. In this dual capacity of Sky-goddess and Cow-goddess, "she exercised her maternal functions from the Gerzean phase of the Predynastic epoch to the beginning of the Roman period, in due course becoming identified with all the local goddesses and the heavens in their entirety, at once the mother and wife of Re and Horus, the 'mistress of the stars', the 'lady of the West' and of the underworld, the goddess of love, of music and the sacred dance."[6]

In Sumer, the great Mother Goddess, Ishtar, was the giver of the increase of the herds and the flocks, and because of this association she assumed the attributes of a cow, frequently being represented with horns in Old Babylonian art (fig. 5).[7] In the form of Ninhursag, the mother of the gods and of men, her temples had sacred herds attached to them. At Lagash, the goddess possessed a "sacred cattle farm,"[8] perhaps the forerunner of the temple goshala in India today. The milking frieze at the temple dedicated to the goddess at Al 'Ubaid shows temple priests preparing and storing the "holy milk of Ninhursag," the nourishment of kings,[9] indicating that in ancient Mesopotamia, as in modern India, milk provided by temple herds fulfilled a symbolic and ritual role in the religious life of the period (fig. 6).

Reverence for the cow as the symbol of the Mother Goddess, her role as both mother and consort to various bull deities, the existence of sacred temple herds, the possible significance attached to milk, perhaps as the life-giving fluid of the Great Mother—all are features of the eastern Mediterranean cattle cults appearing in the Indian cattle complex. This is not to suggest that the origins of the sacred cow concept itself are to be sought in the early riverine civilizations of the ancient Near East. This appears to be a late development in Hinduism in which, superimposed on the traditional Vedic hierarchy of gods and related animals, the cow

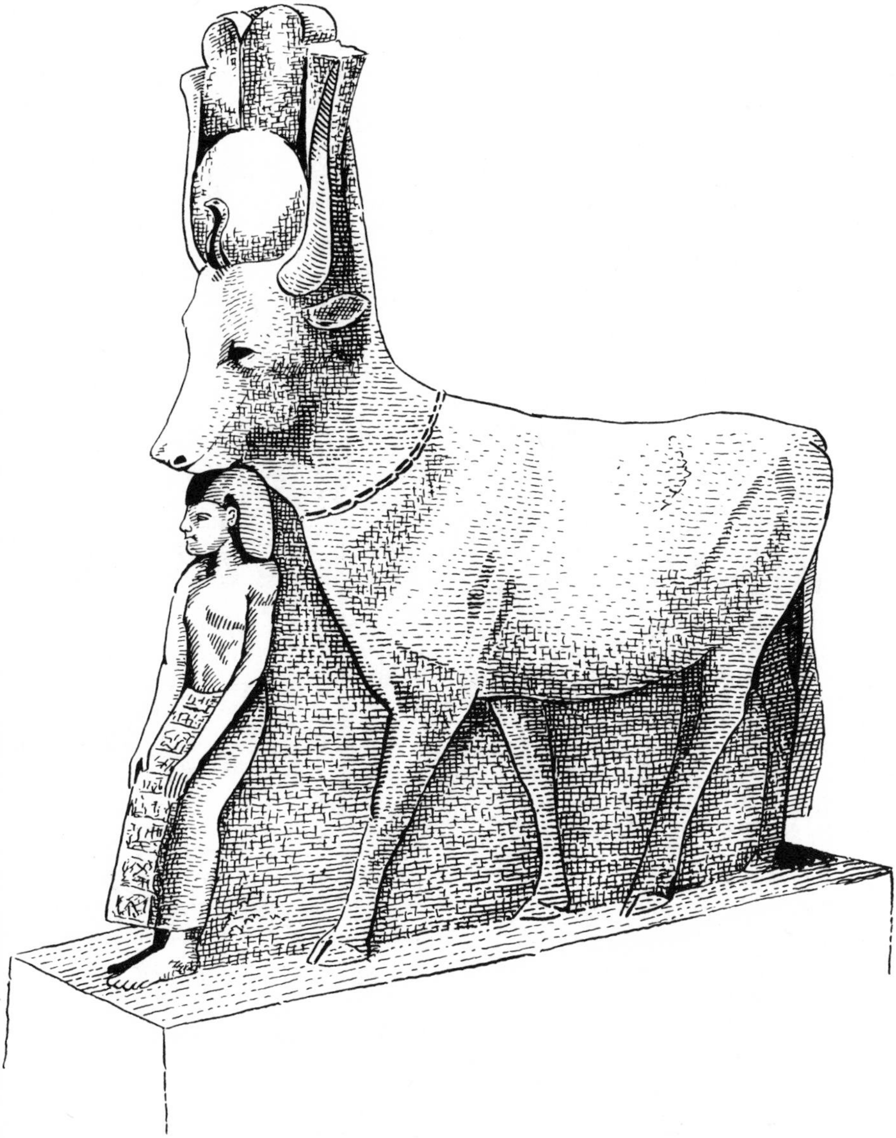

Fig. 4. *The Egyptian Cow-Goddess, Hathor (after Larousse).*

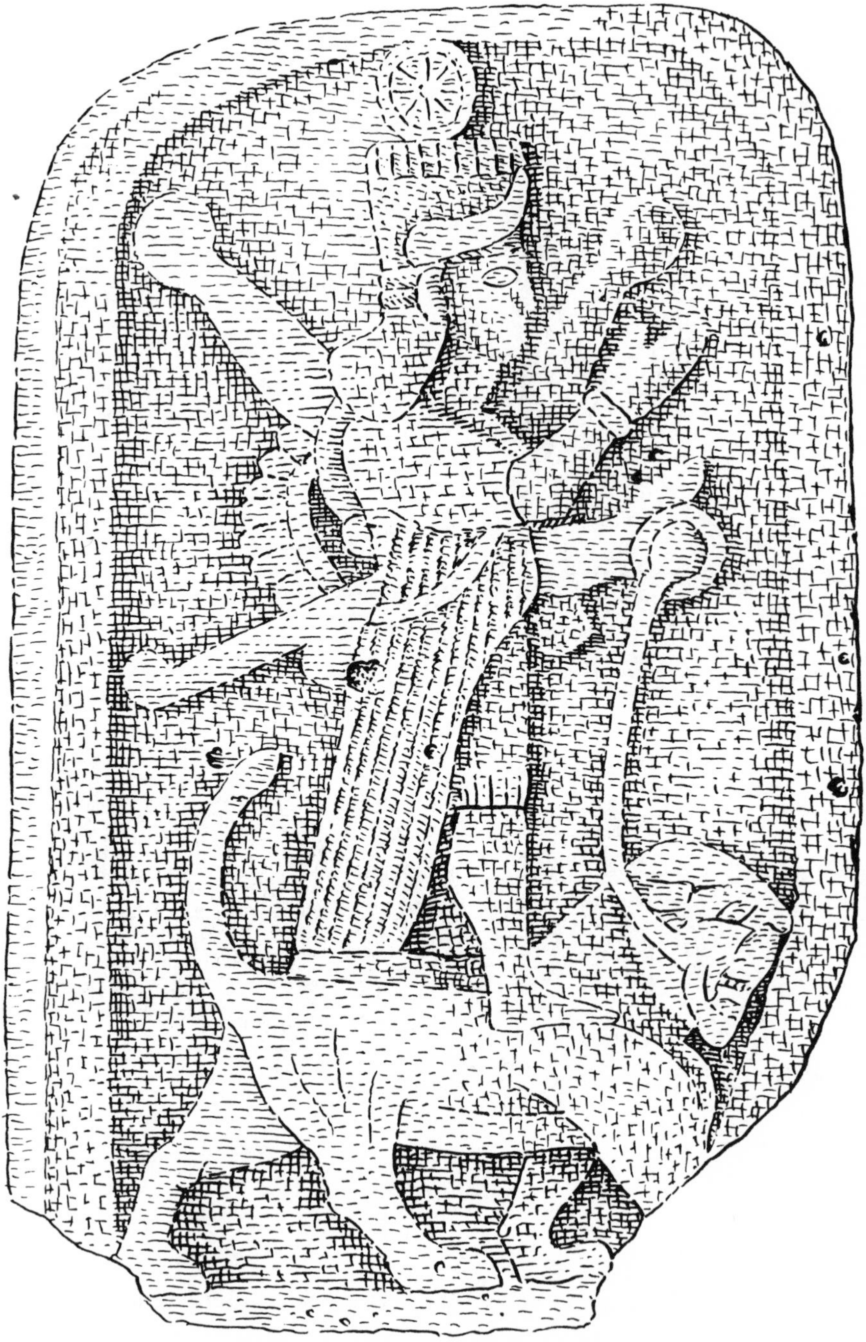

Fig. 5. *Ishtar, the Horned Goddess of Mesopotamia (after Thureau-Dangin and Dunand).*

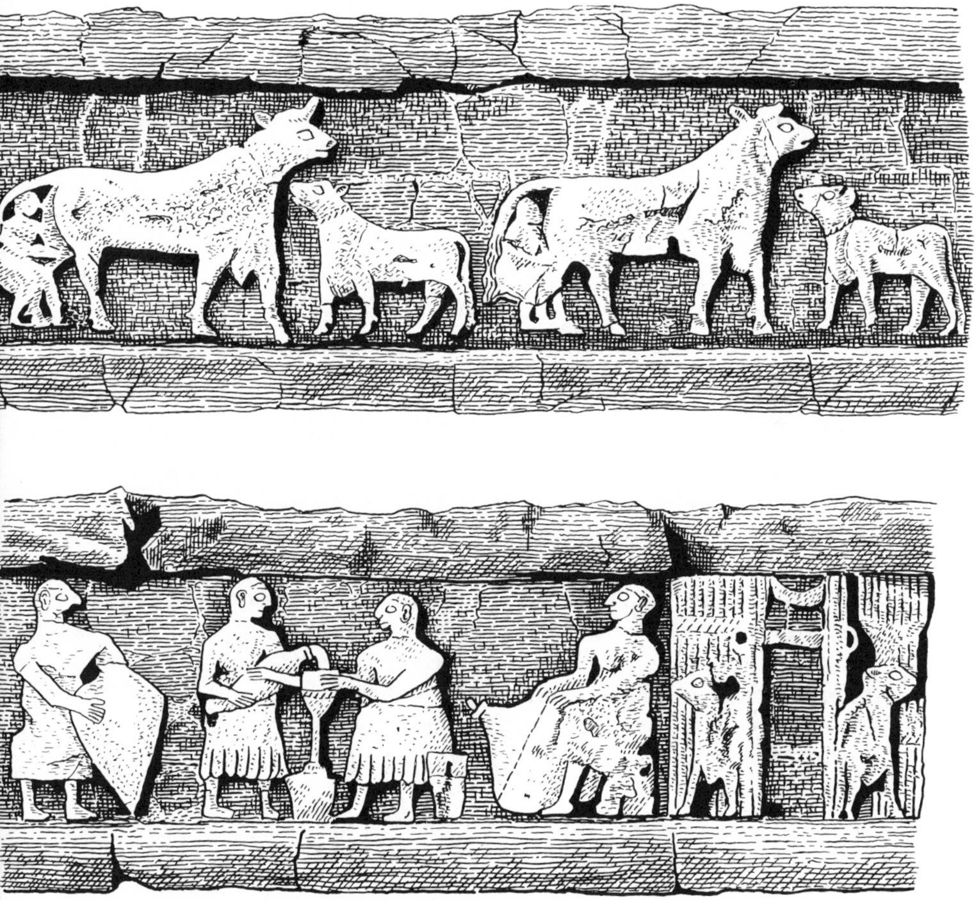

Fig. 6. *Milking Frieze from the Temple of Ninhursag at Al 'Ubaid (after Strommenger).*

emerges as a symbolic focus for a body of metaphysical thought crystallizing in relatively modern times. Despite W. Norman Brown's contention that the doctrine of the sanctity of the cow was "influenced only tangentially by notions originating in non-Aryan cultures,"[10] however, the association of the cow in the Vedas with Aditi, the mother and the wife of the gods, and with the Earth, Pṛithivī, shows remarkable parallels with the Cow-goddesses of the eastern Mediterranean. No matter how this symbolism may have influenced the later development of the sacred cow concept, the survival in modern times of *śakti* worship, of Mother-temples complete with goshalas, of ceremonies such as *Jagadhatri Pūjā* dedicated to the Mother Goddess, and the popular Hindu view of the cow as Mother, all suggest that the cult of the Great Mother, once widespread in pre-Aryan India and with its origins in the ancient Near East, may well be responsible for certain elements of the cattle complex seen in India today.

Cattle in the Indus Valley Civilization

The civilization which appeared in the Indus Valley during the third millennium before Christ shows some remarkable similarities to that of Mesopotamia, and the archaeological record indicates that some form of contact existed between them.[11] As in Mesopotamia, cattle appeared to play an important role in the life of the Indus peoples. At least two species of domesticated cattle, *Bos taurus* and *Bos indicus,* were known, and Marshall cites numerous remains of young animals as evidence for the utilization of cattle as a source of food.[12] This interpretation is supported by Mackay's findings at Mohenjo-Daro and Chanhu-Daro.[13] Prakash regards the presence of domesticated sheep, goats, and cattle as meaning that milk was also an important item of food for the Indus Valley peoples.[14]

Cattle apparently played a major role in the religious life of the Indus Valley civilization. As in Minoan Crete, the cow was conspicuously absent, but the frequency of representation of the bull on seals (fig. 7*a,b*) and as figurines attests to the popularity of the animal. Although the precise nature of the activities centering on the bull is undetermined, the bull and buffalo sacrifice, bull-grappling sports and bullbaiting, the garlanding of bulls and their frequent association with the "manger" symbol suggests some religious function.[15] Basham argues that the bull was sacred,[16] and its occurrence on seals in connection with the enclosure containing the sacred pipal tree (*Ficus religiosa*) and the horned standard of the god indicates, at the very least, a ritual role in Harappan religion. Mackay is of the opinion that the short-horned bull depicted on seals might be a vehicle or emblem of a god of war or of destruction. It is significant that the bull is sacred to Shiva and that the cult of Shiva is pre-Vedic, originating, according to James, in the Harappan civilization.[17] Marshall has identified, on an Indus Valley seal, a Proto-Shiva in the aspect of Pashupati, Lord of the Beasts (fig. 7*c*), although there is question whether this three-faced fertility god is specifically associated with the bull, as is the Shiva of later Hinduism.[18] However, Parpola and his colleagues working on the Indus script have deciphered one of the "fish-signs" of the Indus inscriptions to mean Shiva.[19]

Nowhere is the cow represented in the Indus civilization, although one seal depicts a horned goddess in a pipal tree being worshipped by a horned figure (fig. 7*d*). This deity, resembling the horned Mother Goddesses of Egypt and Mesopotamia, suggests that there was indeed a Mother Goddess cult in the Harappan culture, a contention that is further borne out by the quantity, characteristics, and typical locations of terracotta female figurines recovered from Indus Valley sites. Marshall holds the view that these images probably represented a goddess with attributes similar to those of the "great Mother Goddess." Parpola identifies, on the basis of its frequency of occurrence and location, a particular sign of the Indus script to refer to the Mother Goddess, Ammā. Thus, although there is no evidence for the sanctity of the cow in the Indus Valley civilization, the concept was "deeply laid in the antecedents of the Indo-Iranian Goddess cult where the Great Mother was herself the cow giving her milk as the life bestowing agent par excellence in the process of suckling,"[20] and this cult was of major importance in the religious life of the Indus peoples.

Cattle in Vedic India

The doctrine of the sanctity of the cow and patterns of behavior related to the concept are late developments in Hindu religious thought. Yet many of the ideas critical to the doctrine's formulation have their antecedents in the Vedas, the body of religious literature attributed to the Aryan-speaking peoples who conquered and settled northwestern India during the second millennium B.C.

Unlike the urban agricultural civilization of the Indus Valley, to whose decline they possibly contributed, the Aryans were pastoral peoples whose very livelihood depended on their cattle. In the Vedic literature, the cow, the bull, and the ox are mentioned more frequently than any other species of animal. Numerous specialized words distinguish between types of cattle, such as heifer, barren cow, cow that has ceased to bear after having one calf, and large castrated ox. "It is doubtful if any

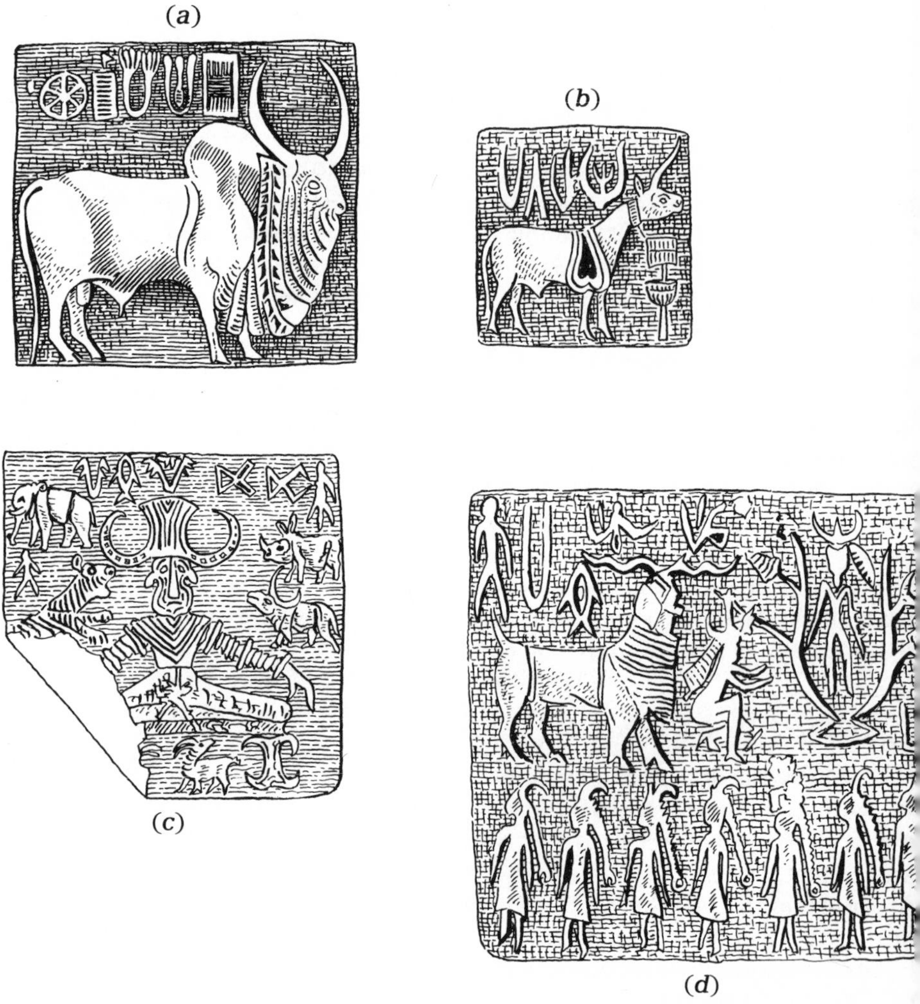

Fig. 7. *Indus Valley Seals (after Marshall, Mackay).*

other large body of literature, belonging to any other people in any period, gives (the) animal so much importance. . . . Cattle or herds of cattle or the products of the cow are the standard items mentioned in descriptions of wealth. Cattle constituted the great booty in war. No other animal was so much appreciated for its economic value."[21]

The importance of cattle in Vedic India was not merely economic, for the animal fulfilled significant ritualistic and symbolic roles in Vedic society. Sacrifice formed the cornerstone of the religion of the Vedas, and cattle were the chief sacrificial victims, their flesh being ritually consumed by priests after the sacrifice. Cow's milk and milk products were used in ceremonies and offered as oblations to the gods. Toward the end of the Vedic period, there are indications that cow dung and ghī were assuming their purificatory role so evident in later Hinduism.[22]

The Vedic literature is also replete with pastoral symbolism.[23] Dyaus, the great Father of the Indo-Aryan pantheon, "the starry Heaven, the bull with the thousand horns,"[24] fertilizes Pṛithivī, the Earth-cow, and from this union spring the gods and all creatures. In the Vedic creation myth, when Indra slays Vṛtra and releases the cosmic waters, they come forth lowing like cows[25] and give birth to the Sun, bringing moisture, warmth, and light into the universe. The Divine Bull, as Indra, is the "seeder who impregnates the cows with his warm *retas* (semen); and as the Sun he is sacrificed for the gain of his warm sap that would nourish vegetation by promoting rain."[26] Parjanya, Soma, Agni, and Rudra are all Vedic deities represented as Bull-gods.

Some of these gods contributed to the emergence of Shiva in later Hinduism and also to Shiva's longstanding association with the bull. The non-Vedic Shiva identified in the Harappan civilization, possibly a fertility deity serving as a consort to the Mother Goddess, and linked with both the bull and the cult of the liṅga (phallus), merged with aspects of Agni and Rudra during the late Vedic period. Subsequently this Shiva-Rudra figure evolved into Shiva, the Great Destroyer in the Hindu pantheon.[27] Today, Shiva is still associated with fertility, being symbolized throughout India by the liṅga and by his *vāhana,* the bull, Nandi. All Shiva temples have statues of Nandi facing the entrance, often with a small shrine dedicated to Nandi himself (fig. 8). The strong ties between Shiva, the bull, and fertility form a historical thread between the bull cults of the ancient world and modern India, and they have undoubtedly contributed to the concept of the sanctity of cattle.

Of equal, if not of greater, importance is the emergence during the Vedic period of the cow as the supreme symbol of femininity in all its aspects of fecundity, maternity, and life-giving sustenance. The cow is Earth; the cow is the mother of the gods; cows are rain clouds; cows are

Fig. 8. *Nandi, the bull of Shiva* (Pallava period, early eighth century A.D. Kanchipuram. Photograph by Helen Desai).

the cosmic waters from which the universe is created. So pervasive is the pastoral symbolism in the Vedas, so widespread the figurative use of the word *cow,* that eventually it comes to be taken literally.

> The equation in Vedic literature and thought of Aditi, Earth and Cow is recognized in the Naighaṇṭuka (1.1.4-5; 2.3.16), where cow (go) is synonymous with many things including earth, heaven, speech (Vac), Aditi. The use of the word or words for "cow" had by then grown from a descriptive figure of speech applied in compliment to feminine entities until it had become a symbol of the holiest of those entities and had finally won identity with them. The metaphor or symbol had run away from those who employed it. They had ceased to distinguish it from the objects it had been meant to adorn or to represent, and thus the cow had acquired their holiness as a quality of its own.[28]

Despite the ritual and symbolic role of cattle in Vedic India, there is no indication that the cow was viewed as sacred and inviolable in its own right during this period. Not only was the cow killed and eaten as part of ritual sacrifice, but meat eating was common in Vedic society. Some authors feel, nonetheless, that evidence within the Vedic literature suggests that the animal was beginning to acquire an aura of sanctity and inviolability. Macdonell and Keith, for example, cite the frequent use in the *Ṛg Veda* and the *Atharva Veda* of the stem *aghnya/aghnyi* (not to be slain), in reference to cattle, as indicative of the animal's sanctity.[29] W. Norman Brown, on the other hand, interprets this as a reflection of the economic usefulness of the animal, noting that the word is used only for cattle that are in milk or carrying a calf, never for useless or barren animals.[30] In another work, Keith quotes instances of prohibition against beef eating in texts such as the *Śatapatha Brāhmaṇa* and the *Kausitaki Brāhmaṇa* as evidence in support of his position,[31] but again Brown suggests that these are not so much restrictions on the eating of meat stemming from considerations of ahiṃsā or the sanctity of the cow, but rather warnings against the improper performing of the sacrificial ritual.[32]

Even if the cow itself were not sacred in Vedic India, there are intimations in the Vedic literature that its presence could impart a degree of sanctity to its surroundings, an attribute that no doubt reflected its value in the economy, its identification with wealth, and its divine associations. Today, although the goshala is certainly not regarded as sacred, it is identified in many ways with religious activities. The temple goshala, of course, is basically a religious institution; but even in the vania goshala the performance of rituals and pūjā, the presence of temples, the celebration of religious festivals, the continued acceptance of the sanctity of the cow and related traditions and folklore, all serve to invest the goshala with a certain religious character. In the Vedas, however, the goshala, or at least its predecessor, the *goṣṭha,* is held to be sacred.

The goṣṭha is not exactly the same as the goshala in the sense of being an asylum for useless cows. Indeed, it would be surprising to find an institution deriving from the sacred cow concept preceding the concept itself. Yet several eminent Sanskrit scholars and religious leaders, including the Śaṅkarāchārya of Puri,[33] with whom the matter was raised, expressed the opinion (or perhaps belief, for lack of any hard evidence) that the goṣṭha of the Vedas could be equated with the goshala of today.

Goṣṭha—literally, "standing place for cows"—refers not so much to cowpens but rather to the pastures or grazing grounds for cattle. "According to the Aitareya Brāhmaṇa (iii,18,14), among the Bharatas the herds in the evening are in the Goṣṭha, at midday in the Saṃgavinī. This passage Sāyana expands by saying that the herds go home to the Shala, or home for animals, at night so far as they consist of animals giving milk, while the others stayed out in the Goṣṭha, or open pasturage."[34] Dange, noting the ritual consecration of the cows before they set out for the goṣṭha as well as the various beliefs concerning the goṣṭha to be found in the *Ṛg Veda,* argues that "it is clearly a very holy place, where health, wealth and nourishment dwell."[35] The idea of the sanctity of the goṣṭha, often translated as cowpen, continues through the Vedic period. The *Vasiśtha Dharma Śāstra* states, "All mountains, all rivers, holy lakes, places of pilgrimage, the dwellings of Rishis, cowpens, and temples of the gods (are) places (which destroy sin)."[36] The *Baudhāyana Dharma Sūtra* forbids one to "enter a burial ground, water, a temple, a cowpen, (or) a place where Brahmanas (sit) without having cleaned one's feet."[37] Other instances of the sanctity of cowpens are found in the *Dharma Śāstras*[38] and even as late as the *Laws of Manu.*[39]

Thus, if the concept of the sanctity and inviolability of the cow is not apparent in Vedic India, there are indications that places where cattle were kept had acquired a degree of sanctity. In addition, there did exist during this period several elements, identified by W. Norman Brown,[40] which contributed to the emergence of the sacred cow concept in later Hinduism. The importance of the cow and its products in Vedic sacrificial ritual, figurative uses of words for the cow in Vedic literature, prohibitions against violations of the Brahman's cow, all played their part in preparing the way for the doctrine of the sanctity and inviolability of the cow. There are even echoes, in the identification of Aditi, Earth, and the cow, of the ancient pre-Aryan association of the cow with the great Mother Goddess, and corresponding links exist between the bull and male deities. But one element that seems to have been critical to the later acceptance of the sacred cow concept does not appear until the end of the

Vedic period, and it may well be of non-Vedic origin. Revulsion against sacrifice, the economic usefulness of cattle, and religious symbolism all were factors contributing to the formulation of the sacred cow doctrine, but it was ahiṃsā that provided the moral and ethical compulsion for that doctrine's widespread acceptance in later Indian religious thought and social behavior.

Ahiṃsā

The doctrine of ahiṃsā, expressing Indian conceptions of the fundamental unity of all life, stands among the foremost virtues espoused by Hinduism, Buddhism, and Jainism. It is seen by many as being India's greatest contribution to ethics and it is manifested in all aspects of Indian life.[41] Some authors argue that, although ahiṃsā is not discussed as an independent and important doctrine, it is not totally absent from the early Vedic writings. It is generally accepted, however, that the first textual reference to the doctrine occurs at the very end of the Vedic period in the *Chāndogya Upaniṣad,* a work belonging to the *Sama Veda* and one which has "contributed the most important materials to what may be called the orthodox philosophy of India, the *Vedānta,* i.e. the end, the purpose, the highest object of the Veda."[42] In a rather obscure passage, listing the virtues appropriate to be a priest's gift, is written "austerity, almsgiving, uprightness, harmlessness (ahiṃsā), truthfulness—these are one's gifts for the priests."[43] The concept appears nowhere else in the Vedic literature, its absence combined with the apparently widespread practices of sacrifice and meat eating suggesting non-Vedic origins. In fact, it may well be that ahiṃsā represents a revolt against Vedic society and its customs. Weber feels that "without question the Jain principle of ahiṃsā originated in the rejection of the meat sacrifice which the Brahmans had illogically preserved out of ancient Vedic sacrificial ritual."[44] Another view contends that ahiṃsā might reflect social tensions or class struggles. By negating the need for sacrifices, both ascetics and Kṣatriyas were undermining Brahman religious and social domination.

Whether or not the ahiṃsā concept was used to challenge the established Vedic order, the origins of the doctrine are not to be found in the Vedas and could even be non-Aryan. Ahiṃsā, along with ideas such as rebirth and *karma,* is not found among any other peoples that share common Aryan linguistic origins,[45] and Sircar notes that some of the early sects favoring ahiṃsā flourished among Aryanized Mongoloid peoples.[46] It is perhaps significant that Bihar, the traditional home of some of the most ardent adherents of the doctrine (Pārśvanātha, Mahāvīra, Buddha) was a region that felt the impact of Aryan civilization at a relatively late date and where, as a result, Brahmanical control was far from complete.[47]

Whatever the origin of ahiṃsā, its first appearance in India coincides with a period of social upheaval expressed, in part, by religious ferment and change. Problems in dating the early Upaniṣads make it difficult to provide an accurate date for the *Chāndogya Upaniṣad* and the first textual reference to the concept. Some authors would place the bulk of the Upaniṣads in the eighth to sixth centuries B.C.,[48] while others claim that on available evidence the earliest acceptable date is the sixth century B.C.[49] This, nonetheless, was an era characterized by popular disaffection with the traditional Vedic religion and the degeneration of the old Vedic order. The spiritual discontent of the times saw the birth of asceticism in Indian philosophic thought, an asceticism that "was not merely a means of escape from an unhappy and unsatisfying world . . . [but] was in part inspired by a desire for knowledge which the four Vedas could not give."[50] The country was full of wandering hermits and ascetics, men who disassociated themselves from the Brahmanical establishment, who reexamined the underlying spiritual values of the Vedas, who spent their days in meditation and reflection on religion, philosophy, and ethics. Out of this seemingly uncoordinated movement emerged the Upaniṣads, philosophical treatises elaborating upon the content of the Vedas and introducing new concepts such as reincarnation, transmigration, karma, and ahiṃsā. Also appearing at this time were the great challenges to orthodox Hinduism—Buddhism and Jainism.

In the centuries following the Upaniṣadic age, from the fifth century B.C. onward, the doctrine of ahiṃsā slowly found favor in Brahmanical circles, so it is mentioned more frequently in Hindu literature. Yet attitudes toward the concept, as far as can be determined from the textual context, remained equivocal and ambivalent. The word is mentioned four times, for example, in the *Bhagavad Gītā* (ca. A.D. 300), but in each case there is no elaboration of the doctrine, which merely appears in listings of human characteristics or virtues.[51] Similarly, in other works such as the *Laws of Manu* (?600 B.C. to A.D. 300?) and Kauṭilya's *Arthaśāstra* (?290 B.C. to A.D. 300?), passages can be found commanding ahiṃsā, while others casually refer to the eating of meat and the slaughtering of cattle. The concept was, therefore, known in Hindu literature, but its widespread acceptance remains doubtful, reflecting considerable discrepancy between Brahmanical rule and popular practice. Even in late medieval India (ca. seventh century), the doctrine of ahiṃsā and related concepts of the sanctity of the cow were still confronting popular apathy and resistance.

Thus it is not in Hinduism that one must seek the initial origins of the animal home in India but rather in those religions that accepted without reservation the ahiṃsā doctrine, Buddhism and especially, as indicated by its association with the modern institution, Jainism.

The Origins of Animal Homes

The earliest documentary evidence for the existence of animal homes in India occurs during the reign of Ashoka (ca. 269–232 B.C.), the Mauryan ruler whose empire covered the entire Indian subcontinent with the exception of a few states in the extreme south of the peninsula. Ashoka adopted Buddhism as the imperial state religion, and the basic tenets of the religion were proclaimed throughout the empire. Several of these official proclamations were engraved on stone pillars scattered around the country, surviving for posterity. Two of these, Rock Edict I and Pillar Edict V, specifically prohibit the slaughter of animals.[52] It is perhaps significant that the long lists of protected animals, though including "bulls set free"[53] and "all quadrupeds which are useful or edible," make no mention of the cow. Of greater interest, however, is Rock Edict II, which states:

> Everywhere in the dominions of King Priyadarśī [Ashoka] as well as in the border territories of the Choḷas, the Pāṇḍyas, the Satiyaputra, the Keralaputra [all in the southern tip of India], the Ceylonese, the Yōna [Greek] king named Antiochos, and those kings who are neighbours of Antiochos—everywhere provision has been made for two kinds of medical treatment, treatment for men and for animals. Medicinal herbs, suitable for men and animals, have been imported and planted wherever they were not previously available. Also, where roots and fruits were lacking, they have been imported and planted. Wells have been dug and trees planted along the roads for the use of men and animals.[54]

The edict arises primarily from a concern for dharma and makes no specific note of animal homes or animal hospitals per se, yet Bhandarkar, Smith, and others accept this to mean that pinjrapoles were in existence during Ashoka's reign. Smith writes, "Although the word hospitals does not occur in the edicts, such institutions must have been included in his arrangements. . . . The curious animal hospitals which still exist at Surat and other cities in Western India also may be regarded as survivals of Ashoka's institutions."[55] One writer, in a discussion of Aryan veterinary science, goes so far as to assert that Buddha himself "established hospitals for men and beasts all over the country, and the institution of Pinjrapoles (Animal hospitals), so peculiar to India, owes its origin to him."[56]

Be this as it may, it is reasonable to accept the existence of the pinjrapole in Ashoka's time, that is, the middle of the third century B.C. It is also quite possible that the institution or something similar was in existence prior to this time, but that it only received official sanction during Ashoka's reign. What is surprising is that the institution should first appear in a Buddhist rather than a Jain context. This is, perhaps, understandable in view of the nature of the evidence available and the relative position of Buddhism (as the state religion) vis-à-vis Jainism in India at the time. Yet all our evidence from present-day India points to the pinjrapole being a Jain institution (see chap. 5).

Admittedly, Buddhism succeeded in eliminating Brahmanical sacrifice in the name of dharma. The animal protection laws of Ashoka, also in the cause of dharma, were as sweeping as any such state legislation in all history. The *abhayāraṇya* (protected forest reserves) of Kauṭilya's *Arthaśāstra,* in which a person who "entraps, kills, or molests deer, bison, birds, and fish . . . shall be punished with the highest amercement [penalties]," foreshadowed the modern game preserve by more than two thousand years.[57] But Buddhism, though the major vehicle for the spread of the ahiṃsā concept throughout India and indeed throughout much of Asia, never carried the doctrine to the extremes of Jainism. In Buddhist thinking, ahiṃsā became a positive adjunct of moral conduct stemming from the cardinal virtue of compassion, rather than the all-encompassing negative principle of nonactivity of the Jains.

The concept of ahiṃsā appears in Jaina philosophy at an early date, possibly even preceding its first mention in the Upaniṣads. Though many scholars believe that Jainism is an offshoot of Brahmanism, the Jains themselves firmly hold the view that their religion antedates the Vedas, perhaps extending back to the Indus Valley civilization. The swastika, an ancient Jain symbol, is frequently found on the Indus seals along with other signs associated with the Jain religion.[58] Jaina tradition has it that the great Jain leader, Mahāvīra (599-527 B.C.), a contemporary of Buddha, was but the twenty-fourth in a line of teachers (tīrthaṅkara) reaching back into the very beginnings of Indian history. The earliest historically substantiated of these, the twenty-third tīrthaṅkara, Pārśvanātha, is thought to have been born around 817 B.C. in eastern Uttar Pradesh. At the age of thirty, he renounced the world and became head of a sizable community of ascetics bound by four vows: not to lie, not to steal, not to own property, and not to kill (ahiṃsā).[59] If any reliance can be placed on the dates for Pārśvanātha, it would seem that ahiṃsā was being practiced among Jain communities before Mahāvīra assigned the doctrine a prominent position in his teachings, and well before the time of Buddha.

Thus, although available evidence places the first pinjrapoles in the reign of the Buddhist, Ashoka, the institutions could well have existed among the Jains at an earlier time. The persistent Jain, and apparent lack of Buddhist, involvement with the institution in modern India, the central position of ahiṃsā in Jain philosophy, and its appearance in Jainism as early as the beginning of the eighth century B.C would all tend to support this point of view.

If Ashoka assiduously promoted the protection of animals in the name of ahiṃsā and institutionalized the pinjrapole, there is little in the Buddhist sources of the time to suggest that the cow was singled out for special treatment. The animal is conspicuously absent from the lists of protected animals, though the bull is mentioned, and beef eating was apparently quite common. Another important text of the same period, the *Arthaśāstra,*[60] elaborates on the office of the Superintendent of the Slaughterhouse, whose duties included the regulation of the sale of meat by cattle butchers (*goghātaka*).[61] Cowherds were permitted to sell flesh, and animals were classified into herds that included "cattle that are fit only for the supply of flesh."[62]

Yet though ahiṃsā was not applied specifically to the cow and though the animal was not viewed as inviolable, the *Arthaśāstra* did impose certain restrictions on the injury and slaughter of some cattle. "Cattle such as a calf, a bull, or a milch cow shall not be slaughtered. He who slaughters or tortures them to death shall be fined 50 panas."[63] Owners of trespassing cattle could be fined and the guilty animals driven off, but they were not to be hurt or killed.[64] But most important of all is the discussion in the *Arthaśāstra* of the duties of the Superintendent of Cows *(Godyaksa),* for these include the supervision of useless and abandoned herds *(bhagnotsṛshṭakam).* "When those who rear a hundred heads made up of equal numbers each of afflicted cattle, crippled cattle, cattle that cannot be milked by anyone but the accustomed person, cattle that are not easily milked, and cattle that kill their own calves, give in return (to the owner) a share in dairy produce, it is termed 'useless and abandoned herd.' " Further, cowherds were directed to apply remedies to calves, and to aged and diseased cows.[65]

Perhaps as early as the fourth century B.C., therefore, herds of useless and abandoned cattle were being maintained in India and warranted mention in Kauṭilya's discussion of animal husbandry practices. It is doubtful that the goshala existed as an institution at this time,[66] but the modern goshala and gosadan must surely be regarded as successors to this ancient tradition of keeping alive unproductive cattle rather than disposing of them. The reasons for this practice are somewhat unclear, but they certainly go beyond the economic and environmental pressures advanced by cultural ecologists to explain the sacred cow concept in modern times. Simplistic arguments relating to the maintaining of cattle numbers for agricultural purposes or for dung production can have little relevance at a time when the population-resource balance was presumably much more favorable than at present. It can be speculated, however, that a complex transformation of society was under way, a process by which certain beliefs were gradually being accepted by segments of the Indian population and were beginning to find expression in their behavior patterns. Thus, despite the ambivalence in the literature of the time toward both the ahiṃsā concept and the sanctity and inviolability of the cow, official sanction of ahiṃsā, increasing restrictions on slaughter, and traditional Vedic attitudes toward cattle were creating an environment, especially among the agricultural classes, which favored the protection and preservation of cattle beyond their useful lives.

From Ashoka to Alā–ud–dīn

We are fortunate to have such sources as the edicts of Ashoka and Kauṭilya's *Arthaśāstra* to shed some light on the beginning of animal homes in India, for the historical record in pre-European times is scanty, indeed. No mention of the institutions is to be found in the Greek, Chinese, or Arab travelers' accounts of India, although the Chinese monk, Fa-hsien, does comment on the presence of charitable hospitals for people run by Vaiśyas in Pataliputra (Patna), formerly Ashoka's capital, at the beginning of the fifth century.[67] Even indigenous sources tell us little of animal homes during this period although, as will be seen, cultural conditions would seem to have favored their existence.

Despite the fact that pinjrapoles may well have existed throughout Ashoka's empire, few traces of the institution exist between Ashoka's time and the arrival of the Europeans in India, though to judge from European travelers' accounts, the pinjrapole was flourishing in western India at the end of the sixteenth century. One interesting piece of evidence, however, does suggest that a tradition of medical treatment for animals continued after Ashoka's reign. A sculpture on a railing pillar from Mathura, belonging to the Kushāna period and dated to the first century, depicts what has been interpreted as a veterinary clinic (fig. 9).[68] The scene shows two monkeys, one with a medical bag slung over its shoulder, apparently operating on a *yakṣa* (a demigod) and a parrot.

Scattered references to goshalas occur, but only toward the end of the

Fig. 9. *Monkey treating bird (from a photograph of the original in the Government Museum, Mathura).*

period in question. In fact, although cow shelters undoubtedly existed before that time, one of the earliest references to goshalas per se relates to an institution established at the end of the twelfth or beginning of the thirteenth centuries. An inscription from south India, dated in the fourth year of the reign of Peruñjiṅga, refers to gifts of cows to a goshala established in the time and in the name of the Choḷa king, Kulottūṅga III (A.D. 1179–1216).[69] Since the giving of cows is a means of acquiring religious merit, this goshala was presumably more than just a cattle farm and it could possibly have been associated with a religious institution.[70] A definite link between goshala and temple is seen in a later inscription (A.D. 1374–1375) on the wall of the Krishnasvāmin shrine in the Padmanābhasvāmin temple at Trivandrum, which tells of the construction of certain buildings, including a goshala, for the worship of the cowherds' god, Krishna.[71]

Despite the relatively late dates of these inscriptions, it seems likely that the goshala appeared well before the twelfth century. Note has already been made of the practice, described in the *Arthaśāstra,* of maintaining "useless and abandoned herds," well before the doctrine of the sanctity and inviolability of the cow had become established. Even though animal sacrifice was revived in the centuries after Ashoka, and even though, in medieval India, "the doctrines of Ahiṃsā and the sanctity of the cow [were] still fighting their way against popular resistance or apathy,"[72] by about the fourth century A.D. both concepts were firmly entrenched in the Brahmanical literature. "Its [the sanctity of the cow] position is made firm doctrinally in Brahmanical circles in the period of the composition of the Purāṇas [ca. fourth to sixth centuries A.D.], and it becomes widely diffused among the Hindu community, gaining ever increasing prestige from that time on."[73] It seems reasonable, therefore, despite the lack of evidence, to place the beginnings of the goshala as a discrete institution in this time.

It is also possible that the pinjrapole, or at least the tradition of the pinjrapole, survived in India from Ashoka's time, despite the fluctuating fortunes of the doctrine of ahiṃsā and of Buddhism and Jainism. Our evidence for animal protection activities during this time is fragmentary, but isolated instances occur of rulers practicing or attempting to enforce ahiṃsā. In the account of his travels in India during the seventh century A.D., for example, the Chinese pilgrim, Hsüan-tsang, writes of the Buddhist ruler, Harsha, "He forbade the slaughter of any living thing or flesh as food throughout the Five Indies on pain of death without pardon."[74] Kalhana, in his *Rājataraṁgiṇī,* tells of a Kashmiri king of the sixth century A.D., Gopaditya, who did not tolerate the killing of animals except for sacrifice, and of Meghavāhana, another Kashmiri king, "who hated killing like a Jaina" and in whose reign effigies of animals in ghī and in pastry were used for sacrifice instead of the animals themselves.[75]

If Buddhism gradually lost ground in India after Ashoka, its rival, Jainism, which also espoused ahiṃsā, gained in strength. Ashoka himself and his successors are thought to have been sympathetic to the religion, and by the beginning of the Christian Era it was well entrenched in Orissa and the northwest. Jainism achieved its greatest prominence in India from the fifth to the thirteenth centuries, spreading throughout most of the country, but it was in the west, under the aegis of the kings of Gujarat, that it indelibly stamped its imprint on succeeding ages, an imprint that survives to this very day in Gujarati culture. Under the royal patronage and favor of rulers like Maṇḍalika, an eleventh-century king of Saurashtra (A.D. 1059–?), Siddharāja Jayasiṃha, King of Gujarat (A.D. 1094–1125) and his successor Kumārapāla (A.D. 1125–1159), Jainism flourished and prospered. Kumārapāla, under the influence of the Jaina monk, Hemachandra, is said to have established Jainism as the state religion and to have assiduously promoted adherence to the ahiṃsā concept. Epigraphic sources indicate that the killing of animals was prohibited throughout Kumārapāla's kingdom and also in the territory of his feudatory princes in southern Mewar, and he apparently set up a special court at Anhilwada to deal with violators of his ordinances concerning the sanctity of animal life.[76] Even if the veracity of the accounts of the dispossession of a merchant for killing a louse or the execution of an offender who had brought raw meat into the capital are open to question, they are indicative of the tenor of life in Kumārapāla's reign. Under these conditions, it is possible that the pinjrapole could have existed, and it is no mere accident that today Gujarat remains at the center of the distribution of the institution over the subcontinent.

The decline of Jainism in the west, first under pressure from Hinduism and then under the onslaught of Islam, marks a turning point in the history of animal protection in India. The depredations of Alā-ud-dīn, who conquered Gujarat at the end of the thirteenth century A.D., did much to destroy the influence of Jainism, and never again was ahiṃsā to be proclaimed as official state policy in a kingdom of India. Even Akbar's *farman* prohibiting animal slaughter, given under the influence of Jaina monks at his court, was directed mainly toward the cow. In the ensuing centuries of Moslem domination in India, it is the protection of the cow that must hold our attention, for herein lie the seeds of an animosity between Hindu and Moslem that has persisted until the present time.

Islam and the Cow in India

The doctrine of the sanctity and inviolability of the cow in India, though well entrenched in Hindu society long before the Moslem invasions of the eleventh century, received added impetus from the introduction of Islamic culture into India. Earlier contacts with Moslems had been limited in extent and generally peaceful in nature, but the military conquests commencing with Mahmud of Ghazni's repeated incursions into the northwest and the subsequent establishment of Moslem political control over Hindu kingdoms introduced an element of conflict into the Indian cultural scene. The Moslem invaders were eaters of beef and killers of cows, and thus the sanctity of the cow came to be regarded as a symbol of Hindu culture, a rallying point for Hindu resistance against the spread of Islam.

Not all Moslems ignored the traditional Hindu attitudes toward the cow, and several rulers, as a matter of statesmanship, promulgated anti-slaughter legislation. Babar (A.D. 1526–1530) is supposed to have issued a farman forbidding cow slaughter, and Akbar's injunction against the slaughter of animals has already been noted. The Frenchman, Bernier, writes that Akbar's son, Jehangir, "at the request of the *Brahmens,* issued an edict to forbid the killing of beasts of pasture for a certain number of years."[77] One Moslem writer suggests that instead of issuing a general order prohibiting cow slaughter, the early Moslem rulers imposed a *jazri* tax on butchers slaughtering cows to accomplish the same ends,[78] though this interpretation is somewhat suspect.

With perhaps the exception of Akbar, whose eclecticism and catholic tastes in religious matters led many orthodox Moslems to view him as apostate, Moslem efforts at cow protection were motivated not by any reverence for the sanctity of life or for the cow in particular, but were seen rather as a means of ensuring the good will of the Hindu population. In his will, Babar instructed his son and successor Humayan, "Avoid especially the sacrifice of the cow by which thou canst capture the hearts of the peoples of India, and subjects of this country may be bound up with royal obligations."[79]

Yet, despite these sentiments and the various restrictions on cow slaughter imposed by the rulers of Moslem India, it is clear that the killing of cows was a very real issue between the Hindu and Moslem communities, one that has continued to cause tensions even in India today. The slaughter of cows, the presence of Moslem butchers, the eating of beef, all offended the sensibilities of the orthodox Hindu and were used by the Brahmans to generate Hindu feeling against Islamic religion and culture. The Moslem festival of *Bakr Īd*, at which cattle were sacrificed, was the cause of much unrest in the Hindu community. Even today it is the occasion of special donations to pinjrapoles for the liberation of animals destined for the sacrificial knife. Furthermore, if the doctrine of the sanctity of the cow came to be strengthened by identification with Hindu culture, violations of the doctrine were seen to be a means of encouraging the spread of Islam. Mukandi Lal, for example, argues that Moslems encouraged Hindus to eat beef so that they would thereby become outcaste from Hindu society and turn to the religion of Allah.[80] In this context, it is perhaps significant that Moslem proselytizing was most effective in the lower levels of the Hindu caste structure, where the sanctity of the cow was less well entrenched and where beef eating was not uncommon.

The use of the cow as a cultural and, indeed, political symbol of Hindu resistance to Moslem power in India achived its greatest success during the Maratha struggle against Moghul rule during the seventeenth century. For Shivaji (A.D. 1627–1680), the leader of the Maratha revival in the western Deccan and founder of a Hindu Empire that survived until the British period, his sacred duty was to protect the cow and the Brahman from the Moslem. "We are Hindus and the rightful lords of the realm. It is not proper for us to witness cow slaughter and the oppression of the Brahmans."[81] He is popularly viewed by the Hindus as an incarnation of God who assumed human form to deliver the cow and the Brahman, those two pillars of the Hindu faith, from the hands of the Moslem in India.[82] Under Shivaji and his successors, the sanctity of the cow became a standard of the renascent Hindu power in western India, an area which to this day remains a stronghold of conservative Hinduism and in which the doctrine of the sanctity of the cow is well entrenched. The survival of the goshala and the pinjrapole in western India no doubt owes much to the Maratha rulers of the seventeenth and eighteenth centuries.

The Cult of Krishna

The Moslem presence in India did much to promote, albeit indirectly, the doctrine of the sanctity of the cow in Hindu society. Even today, Hindus are known to explain the establishment of their goshalas in terms of protecting the cow from the depredations of the Moslems. Of far greater significance, however, is the role that the cult of Krishna has played in fostering the respect and reverence many Hindus have for the cow.

Krishna is by far the most important of the *avatārs* (incarnations) of Vishnu and one whose popularity has surpassed that of the more ancient Vedic deity. Vishnu is the preserver of the world, but it is Krishna, the great hero, Krishna, the Divine Cowherd, Krishna, the mischievous Child-god who has been the great inspiration in much of Indian art, literature, poetry, and philosophy. The *bhakti* (devotional) movements of medieval India attained their greatest heights in the worship of Krishna, and even today his message of love and devotion is being carried into the western world by the members of the Hare Krishna sect.

Although over the centuries many elements of diverse origins have been brought together in the Krishna myth, most scholars recognize three distinct aspects of the god: the God-hero; the divine child; and the pastoral Krishna. A Krishna Devakiputa (Krishna, son of Devaki) is mentioned in the *Chāndogya Upaniṣad,*[83] and Raychaudhuri identifies this Krishna with the Krishna of the later epics.[84] In this form, Krishna becomes the great God-hero of the *Mahābhārata*, the divine companion and guide of Arjuna in the *Bhagavad Gītā*, the heroic God-king who meets his death after the internecine slaughter of the Yādavas at Dwarka. In total contrast to this majestic figure of Indian mythology is the plump and appealing Child-god, Krishna, so close to the heart of Indian womanhood. A late addition to the Krishna legend, perhaps inspired in part by tales of another God-child carried east by Christian merchants and missionaries,[85] this dimension of the divine child holds a tremendous popular attraction in a society traditionally viewing the prime role of women to be the bearing of male children. But of far greater relevance for our purposes is the third aspect of the deity, Krishna in the form of Krishna Govinda, the Lord of the Herdsmen.

The pastoral element in the cult of Krishna differs in origin from that of the god as hero or child. In the Vedic literature, it is Vishnu, not Krishna, who is presented as "protector of cows" *(gopa)* and "chief herdsman" *(govinda).*[86] Krishna, on the other hand, first appears in his pastoral aspect in southern India, and it has been suggested that in this form he may originally have been a fertility god of the peninsula, whose cult was carried northward by nomadic pastoral peoples. The Ābhīras, a nomadic tribe of professional herders who appeared in southern Rajasthan, Malwa, and the Sind around the beginning of the Christian Era, are thought to have been important in spreading the worship of the pastoral Krishna.[87] This view is challenged, however, by Majumdar, who argues that Krishna worship must have originated with the Gopas of Vraj, the legendary homeland of Krishna, in the direct tradition of the Vedic hero rather than through derivation from a non-Vedic tribal deity.[88]

It is quite likely that both Vedic and non-Vedic elements were combined in the cult of Krishna appearing in the Mathura region under the Kushāns (A.D. 65–225).[89]

Krishna and the cow are inseparable in the Hindu mind. Works such as the *Harivaṃśa Purāṇa* and the *Bhāgavata Purāṇa* tell of the pastoral nature of Krishna's early life among the Gopas of Vraj, and Krishna himself says, "We are cowherds, wandering in the forests, maintaining ourselves on cows, which are our wealth; cows are our deities, and mountains and forests."[90] Krishna is further identified with pastoral people in incidents such as that commemorated by *Goverdhan Pūjā,* when the god sheltered his people and their herds from Indra's rains by lifting Goverdhan Mountain—an event which is interpreted by some as reflecting conflict between rival pastoral and agricultural economies. It is not surprising, therefore, to find the cult of Krishna well entrenched among the pastoral castes in India today. Cattle-keeping groups, such as the Āhīrs and the Goālas, are involved the most actively in celebrating Krishna-related festivals, such as Goverdhan Pūjā, *Gopāṣṭamī,* and *Janamaṣṭamī.*[91]

The worship of Krishna is not, however, restricted to pastoral peoples in India, and Krishna is one of the most popular deities of the Hindu pantheon. As an incarnation of Vishnu, he is revered by all orthodox Vaishnavas, who are usually strict vegetarians and ardent worshippers of the cow. Though much of the mass appeal of both Krishna and the cult of the cow dates back to the fifteenth century revival of Vaishnavism, which saw the translation of the *Bhāgavata Purāṇa* into Hindi, the language of the common people, Krishna worship has long been a feature associated with the Vaiśya *varna* of the Hindu caste system, the traders, merchants, and agricultural classes.[92] Vāsudeva-Krishna himself is said to be descended from a Vaiśya group,[93] and the links between Krishna, the Vaiśya, and the cow are further reinforced by the *Bhagavad Gītā's* injunction to the Vaiśya to protect cattle.[94] This cultural trilogy achieves its greatest expression today in the Pushti Margi cult of Krishna worshippers. Founded by Vallabhāchārya, in whose teachings reverence for the cow plays a significant role and who is chiefly responsible for popularizing the cult of the cow in recent centuries, the Pushti Marg, "The Way of Grace," finds its strongest support among the vania castes of western India. It is among these Vaishnavas of Gujarat and Rajasthan, followers of Vallabhāchārya, that the worship of the cow attains a prominence scarcely equaled elsewhere in India. It is by no means coincidental that the goshala in India today is essentially a Vaishnava institution.

The European Period

If there is evidence that the temple goshala existed in India by the twelfth century A.D., it can only be surmised that the pinjrapole survived from Ashoka's time. European travelers' accounts from the late sixteenth century onward, however, show that the institutions were flourishing, not only in western India but in other parts of the country as well. Mention of them is first made by Ralph Fitch, the English merchant who traveled in India between 1583 and 1591. In addition to his comment on the animal hospitals in Cambay, he writes of his visit to Cooch Behar, at present part of eastern Jalpaiguri District in West Bengal, "Here they be all Gentiles and they will kill nothing. They have hospitals for Sheepe, Goates, Dogs, Cats, Birds and for all other living creatures. When they be olde and lame, they keepe them until they die. It a man catch or buy any quicke thing in other places and bring it thither, they will give him money for it or other victuals, and keepe it in their Hospitals or let it go."[95]

All the seventeenth-century travelers remark on the pinjrapoles of Gujarat, run mainly by the Jains and the "Banians." Ovington provides us with a description of a hospital at Surat built for "the preservation of Buggs, Fleas, and other Vermin, which suck the Blood of Men; and therefore to maintain them with that choice Diet to which they are used, and to feed them with their proper Fare, a poor Man is hired now and then to rest all Night upon the Cot, or Bed, where the Vermin are put, and fasten'd upon it, lest the stinging of them might force him to take his flight before the Morning, and so they nourish themselves by sucking his Blood, and feeding on his Carcass."[96] Fryer writes of hospitals for cows at Tanore, in Kerala, which could easily have been goshalas rather than pinjrapoles.[97] Tavernier is supposed to have seen a similar hospital for monkeys at Mathura.[98] Manucci, Mundy, Thévenot, and many other Europeans in India at this time refer to what they considered these rather unusual institutions of the Hindus in the accounts of their travels.[99]

From the eighteenth and nineteenth centuries, more detailed descriptions of pinjrapoles and their workings are available, especially in the official Gazetteers of the Bombay Presidency, but none is more vivid that that of the Frenchman Rousselet:

> One enters first into a large courtyard, surrounded by sheds, in the middle of which are a hundred oxen. There is nothing more curious than this gathering of sick quadrupeds: some have bandages over their eyes; others, lame or in a helpless condition, are comfortably stretched out on fresh straw. These animals appear to have something human about them; they call to mind a hospital scene by Granville. Servants wash them, rub them down and bring the blind and paralyzed their food. In this courtyard, there are also horses, asses, dogs and cats.

> Some of these animals appear to be so sick that I venture to tell my guide it would be more charitable to put an end to their suffering. "But," he replies, "is that how you treat your invalids?" A little farther on, I enter an enclosure reserved for bipeds; aged crows that have committed all manners of crimes live out their lives peacefully in this paradise of beasts, in the company of bald vultures and buzzards that have lost their plumage. At the end of the court, a heron, proud of his wooden leg struts about in the midst of blind ducks and lame fowl. All the domestic animals and all those that dwell in the vicinity of mankind are represented here; rats are seen here in great numbers and display a remarkable tameness; mice, sparrows, peacocks and jackals have their asylum in this hospital.[100]

The pinjrapole described in this account written in the late nineteenth century differs little in its essentials from the institutions of today.

The nineteenth century is also of significance for our discussion of the animal home in modern times, since many pinjrapoles and virtually all vania goshalas existing in India today were established within the last 150 years. This raises the question of why during this period there should have been a sudden upsurge in interest and concern for the sanctity of life in general, and for cow protection in particular. The problem has no simple solution, but it is possible to identify several factors that have contributed to increasing public awareness and involvement with animal homes during the last two centuries.

Undoubtedly economic conditions prevailing during this period are responsible for the existence of many goshalas. Mention has already been made of institutions established to protect cattle during natural calamities such as the Bengal famine of 1891. Others were set up to provide milk to schools, ashrams, and other institutions, or for the purposes of cattle development. Increasing cattle populations and pressure on available resources probably led to greater numbers of useless and stray animals in the economy, creating a need for cattle protection.

Economic factors alone, however, cannot account for the growing popular support for cow protection in India during the late nineteenth and early twentieth centuries. Of much greater import was the confrontation yet again of Hinduism with an alien culture. Just as the sanctity and inviolability of the cow were used as a Hindu symbol to combat the spread of Moslem culture and as a rallying cry for Maratha resistance against Moghul power, so the concepts of ahiṃsā and of the sacred cow were identified with the struggle against Western influence in India. One reason advanced for the emergence of goshalas and pinjrapoles at this time was the imperial government's lack of interest in the protection and preservation of the cow. Although general violations of the ahiṃsā doctrine led to the founding of institutions such as the Bombay and Poona pinjrapoles, the cow soon emerged as the specific focus of Hindu feeling against the foreign rulers of the country. The Indian Mutiny is only one instance of how feelings could be aroused by charges, either real or alleged, that the government was using cow's fat to undermine the religion of its Hindu sepoys. The British predilection for beef did much to further the cause of cow-protection in India, and even though Christianity never posed much of a threat to Hinduism, the eating of beef was very much a sign of conversion to the Christian faith. Educated, westernized Hindus even flaunted their beef eating as evidence of their progressive thinking.

As in Moghul India, the cow and attitudes toward the cow became associated with conflicting cultures, so much so that the traditional Hindu respect and reverence for the cow became a cause célèbre in the nationalist movement. Just as *khādi* and the Gandhi cap identified the wearer as a sympathizer with the nationalist cause, so veneration of the cow became the sine qua non for those who supported independence from the British. *Gomātā,* Mother Cow, became the emblem of "Indianness," and it is not surprising to find today that the symbol of the Congress Party, the successor to the Indian National Congress, is the cow and suckling calf.

It was in this political and social climate that many of the animal homes in India were founded. While their existence cannot be entirely explained in these terms, it is evident that the British presence in India contributed to the resurgence of concern for the sanctity of the cow during the nineteenth and twentieth centuries and acted as a catalyst to the revival of those traditional Hindu values that find expression in goshalas and pinjrapoles. The extent to which the values of ancient India continue to be preserved in the animal shelters of today is seen in the accounts of contemporary institutions presented in the following chapters.

V

Compassion for Life—Case Studies of Pinjrapoles in Modern India

Ahiṃsā is like a loving mother of all beings.
Ahiṃsā is like a stream of nectar in the desert Saṃsāra.
Ahiṃsā is a course of rain-clouds to the forest-fire of suffering.
The best herb of healing for the beings that are tormented by the disease
Called the perpetual return of existence is Ahiṃsā.
—Hemachandra

Although it has been shown that goshalas and pinjrapoles are centers for the preservation and protection of animal life, the manner in which these ends are accomplished and the many implications of the institutions' activities have yet to be explored. Questions must be answered about details of organization, financing, and management, and about the economic and social dimensions of the institutions. To examine these varied aspects of animal homes in India, detailed case studies of selected goshalas and pinjrapoles are undertaken in this and the following chapter. The institutions have been chosen to provide a comprehensive picture of the animal homes of today, their workings, and their relationship to the broader patterns of contemporary Indian society.

The Ahmedabad Pinjrapole Society[1]

On the banks of the Sabarmati River in Gujarat, some eighty kilometers from its mouth in the Gulf of Cambay, lies the city of Ahmedabad. A modern industrial center of over one-and-a-half million in population, Ahmedabad derives most of its current prosperity from its flourishing textile industry. But, unlike other modern industrial cities such as Bombay, Calcutta, or Madras, which owe their very existence to the British, Ahmedabad can trace its history back to A.D. 1411 when:

> Clearing the neighbourhood of robbers and highwaymen, Ahmad[2] built a citadel or fort of much strength and beauty, calling it Bhadar after the Patan [Pathan] capital; laid out his city in broad fair streets and added to the Sabarmati's scanty stream by turning into it the waters of Hathmati. Bringing marble and other rich building materials . . . , he raised magnificent mosques, palaces and tombs, and gathering merchants, weavers and skilled craftsmen, made Ahmedabad a center of trade and manufacture.[3]

Since that time, despite fluctuations in its fortunes, Ahmedabad has always been a commercial and manufacturing city of note in western India.

Much of the preeminence of the city during medieval times was due to its location, for it lay astride the main land routes from the Gujarat ports, the chief of which was Cambay, to the princely courts of Delhi, Rajasthan, Agra, and Malwa. Caravan after caravan of donkeys and camels headed northward through the mountains and deserts of Rajasthan or eastward into the hills of Malwa, laden with goods from Europe and the Middle East, returning with drugs, spices, cloths, and other items for export, manufacture, or exchange. But Ahmedabad was more than just a trade center, for it had gained an international reputation as a producer of textiles. By the beginning of the sixteenth century, the silks, brocades, and cottons of Ahmedabad were in great demand "in every eastern market from Cairo to Pekin" and on the coast of Africa were exchanged for gold often at one hundred times their real value.[4]

In association with the emergence of the city as a commercial and textile center there developed a social and economic infrastructure that set Ahmedabad apart from most Indian cities. The vestiges of this may be seen even today. In a city founded and ruled by Moslems until the coming of the Marathas in the eighteenth century, in which the skilled labor force was Moslem, the wealth of Ahmedabad was concentrated largely in the hands of non-Moslems. The financiers, the bankers, and the traders were generally Hindu or Jain by religion, and it was these powerful mercantile communities that set their stamp on the city. As Gillion comments:

> The influence of the financiers and merchants of Ahmedabad appeared in the survival of certain old institutions of Hindu mercantile cities: the mahajans and the Nagarseth. Whereas in other parts of India the guilds, so powerful in the time

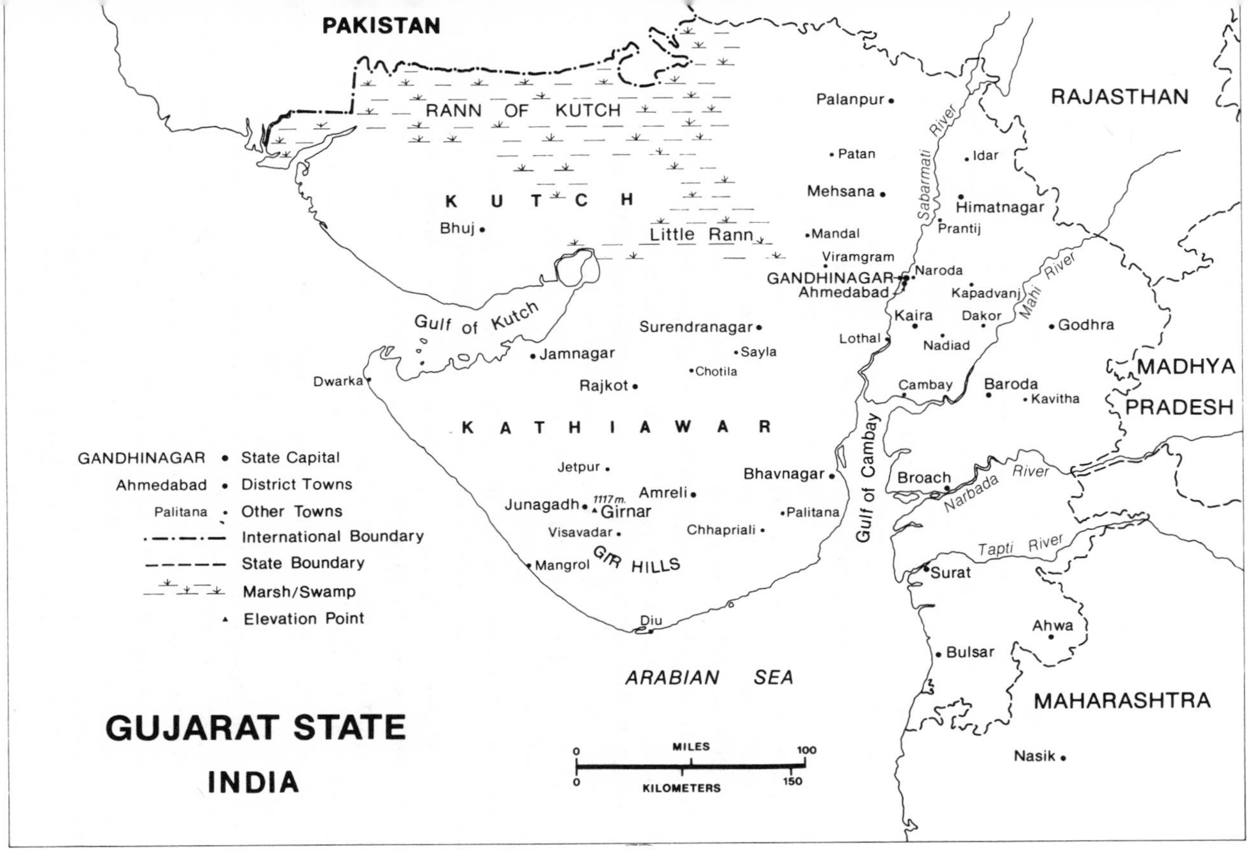

Map 7. *Gujarat State*

> of Buddha and for centuries afterwards, had all but disappeared, squeezed between royal and caste power, they were still important in the nineteenth century in Gujarat, and to a lesser extent in Rajasthan. . . . In Ahmedabad, they are by no means unimportant even now. In Mughal and pre-Mughal days, they were as much rulers of the city as the royal governors and officials and for the individual of far more immediate significance.[5]

There were some forty guilds in Ahmedabad during the nineteenth century, and even today the *mahājan* (guild) survives throughout western India, wielding considerable economic and social power. Traditionally, both the guilds and the *Nagarseth* have been closely associated with the running of pinjrapoles, and these interesting social institutions of ancient India deserve further comment.

In medieval times, the title of Nagarseth, literally "city chief," was accorded to the leading citizen of the mercantile communities of western India. By the middle of the nineteenth century, this term had fallen into disuse in most towns in Gujarat, merely implying that its holder was the chief alderman of the bankers' guild. In Ahmedabad, however, the Nagarseth was undoubtedly the single most important merchant in the city; in addition to being head of the powerful bankers' guild, he was the de facto leader of the city's Jain community, one of the largest and most influential in all of western India. Even today, although the city is run by a bureaucracy headed by a government-appointed Administrator, the *seths* of the city, the leading merchants, exert considerable power in community affairs and local politics.

The word *mahājan,* according to dictionary definitions, can mean "banker" or "illustrious or wealthy man," but in Gujarati it is also used in the sense of a guild or trade association. The jurisdiction of the mahājans is restricted essentially to trade matters, to determining holidays, and to collecting and dispensing corporate funds. They regulate working hours and wages and generally look after the welfare of their members. Where they have been powerful enough, as in Ahmedabad, they have been able to exclude outsiders from the commerce of the city.

Perhaps of greater significance in the past was the power of the mahājan to arbitrate trade disputes and even to impose social sanctions against offending members. Expulsion from a guild was often a social disaster. The Gazetteer of the Bombay Presidency cites the example of a Visa Śrimali vania in Ahmedabad who earned the displeasure of his caste brothers by marrying a widow. As a result, he was expelled from the cloth dealers' mahājan and eventually was forced to close his shop and move to another town.[6] Even today these sanctions may still be imposed. A former president of the Maskati Cloth Market Association, the largest cloth dealers' mahājan in Ahmedabad, reports that if a member were to be expelled from the mahājan for some reason, he would no longer be permitted to rent a shop in the market—a serious blow to his business opportunities in a society in which communal ties are all-important.[7]

As might be expected, the most powerful guilds in Ahmedabad were those associated with the traditional textile industry—the *śarafs'* (bankers') mahājan, the cloth dealers' mahājan, and the raw silk dealers' mahājan. In most instances, institutions such as these would be expected to remain strongholds of resistance to change and innovation; yet, despite their conservative nature, it was these very institutions and the vania communities supporting them that contributed in no small way to the ability of Ahmedabad to make the transition from a traditional industry to a modern, steam-powered, factory-based one. Indeed, the transition has been so successful that the city has earned the accolade, if it can be called that, of being known as the "Manchester of India."[8]

Today this prosperous industrial city has the largest number and greatest variety of animal homes anywhere in India. Within the limits of Ahmedabad are three pinjrapoles, three temple goshalas, four vania institutions, and (formerly) one Gandhian goshala, the last being in the ashram on the banks of the Sabarmati that Gandhi himself founded and lived in during his days in the city (table 1). One of these, the Ahmedabad Pinjrapole Society, is of special interest since its records, the oldest extant records of any institution that I have been able to locate, show it to have been in existence for at least two centuries. It is one of the largest pinjrapoles in the country, and because of its age and location in a city that has preserved much of its medieval character, it will provide a reasonably accurate picture of a pinjrapole in a traditional Indian setting.

No available records name the founder or date the beginning of the Ahmedabad Pinjrapole, but the society has account books dated to the year 1814 of the Vikram Samvat era, corresponding to A.D. 1758. The pinjrapole is, therefore, over 200 years old, and it does not seem unreasonable to argue for an even greater antiquity. Fitch noted the existence of a hospital for animals in Cambay during his journeys in India between 1583 and 1591,[9] a period in which Ahmedabad was approaching the height of its prosperity under the Moghuls and was a great center of trade and commerce. It is plausible that the city would have had its own pinjrapole at this time and, if the modern situation is an accurate reflection of the past, that it would have been founded and supported by the Jain mahājans of Ahmedabad.

The early records of the Ahmedabad Pinjrapole, mostly account books, tell little of the institution beyond its day-to-day functioning. A typical

TABLE 1

GOSHALAS AND PINJRAPOLES IN AHMEDABAD

Type	Institution
1. *Pinjrapole*	Ahmedabad Pinjrapole Society Ahmedabad Khoda Dhor (Helpless Animals) Pinjrapole Manekchowk Khoda Dhor Pinjrapole
2. *Temple Goshala*	Jagannath Mandir Goshala (Krishna Temple) Swami Narayan Mandir Goshala (Krishna Temple) Nilkant Mahadeo Mandir Goshala (Shiva Temple)
3. *Vania Goshala*	Harihar Maharaj Kāmdhenu Gosevashram Gosamvardhan Ashram Goshala Janaki Gosevashram Santosh Gosevashram Public Trust
4. *Gandhian Goshala*	Ashram Goshala, Sabarmati

(Source: Gujarat State Federation of Goshalas and Pinjrapoles.)

page from an account book, dated Samvat 1858 (A.D. 1802), is seen in figure 10. The entry opens with an invocation to the god, Ganesh, the remover of obstacles, followed by the date, the ninth day of Fagan Sud (the bright fortnight of the month of Fagan) in the Samvat year 1858. The first line in the left-hand column repeats the invocation to Ganesh, followed by a statement indicating a cash balance of Rs. 43-8 annas. Next is an entry dated 4 Fagan Sud for a donation of Rs. 10 in Bombay money (money used in British territory as opposed to that of the State of Baroda) from one Sakarchand Mukaidas for the purchasing of feed for the pinjrapole's animals. Following this is a debit item, recording the payment of Rs. 25-12 annas to a milkman, Maoo Jamal Parma, for milk provided to the pinjrapole.

Of much greater interest is the record of a dispute during the early part of the nineteenth century involving the Jain and Vaishnava communities and concerning the dispensation of funds in the city of Ahmedabad. In 1827 the following petition, quoted in full, was submitted to the British Governor of the Bombay Presidency:

To the Hon:
Mountstuart Elphinstone.
President and Governor in Council.

The most respectful Petition of the undersigned Śravak Banians[10] *residing at Bombay and Ahmedabad through their Agents and Gomashtahs [clerks].*

Sheweth—

Your Petitioners beg to bring to the notice of your Honourable Board that from time immemorial a custom existed of collecting on all kinds of Goods and Merchandise at Ahmedabad one quarter per cent exclusive of Customs on the Exports from that City for purposes of Charity such as maintaining and provisioning of Cattle and Animals of every description, a Charity the appelation of which is Panjarapole, strongly enjoined by the Veedas and all other sacred books of the Hindoo religion.

In the time of the Mogul Governments as well as those of the Peishwa and Guicowar,[11] *this custom has always been allowed and protected.*

Upon the British Government assuming its authority there, the respectable Mr. Dunlop, the Collector, also countenanced the charitable Institution being in conformity to ancient usage and the Customs of your Petitioners' Caste religion.

Now may it please your Honourable Board, your Petitioners have lately learnt that it is the intention of the Caste of Misree Banians in future to give this Charity to a man whom they worship and whom they call Maraj—Thus violating one of the most sacred institutions and breaking through all established ancient use and Custom in so religious an affair.

Your Petitioners worship no man and their law does not justify them to depart from the duty of provisioning and maintaining poor and helpless animals.

This principle of tenderness is, however, not confined to the horned species alone: our belief in the metempsychosis makes us extend it to every living creature.

Your Petitioners under the foregoing representations entreat of your Honourable Board to prevent any innovation in this most charitable institution by the Misree Caste of Banians or any other sect, and that as heretofore they may be permitted to recover the duty as usual for the purpose of the above mentioned.

And your Petitioners as in duty bound, shall ever pray—

Bombay
9th July, 1827

Signed by Moteechund Amichund [and two-hundred-and-thirty-four other merchants of Bombay and Ahmedabad].[12]

It would seem to have been a long-established custom for a tax to be levied on business transactions for the maintenance of the pinjrapole, a practice going back to the time of Moghul rule. The Śravak Vanias, Jains, had heard that the Misree Vanias, who were Vaishnava Hindus, intended to donate their share of this tax to the *Mahārāj*, the spiritual head of the Vaishnava community in Ahmedabad, and had petitioned the Governor of the Bombay Presidency to force them to continue to support the pinjrapole.

The Office of the Governor referred the petition to the Collector of Ahmedabad, who reported (No. 21 of 1827, Judicial Department) that even though the practice might have been endorsed by the native rulers of the city, the Śravaks had no *sunand* or deed of grant for their claim to the money. The funds had been collected not by the government but by the clerks of the Śravaks, who then paid the Nagarseth, in whose hands lay the management of the institution. The payment of the tax was the result of a voluntary agreement, as indicated by the following document:

Samvat 1877, Asso Sud 2nd, corresponding with 23rd October, 1820.

We the Mahajans of the city of Ahmedabad, having met together, hereby execute this written agreement engaging ourselves to discharge the long established contributions for the support of the Khoda Dore Pinjrapole as has always been done as also the quarter percent collected and paid for the same purpose on goods purchased, that these dues shall be paid on the spot without any excuse or delay, and any person paying less shall forfeit his religious rights.

The document was signed by representatives of twenty-two mahājans of Ahmedabad.

Fig. 10. *Account Book from Ahmedabad Pinjrapole*

The Vaishnavas countered by presenting a petition on behalf of the "Gadee of Mahārāj or Ahmedabad's inhabitants," claiming that the fee on goods imported and exported through Ahmedabad was granted by the King of Delhi exclusively for the maintenance of the Mahārāj and his family. They produced various documents supporting their case, with one sunand even bearing the signature of Ahmad Shah, founder of the city, and they requested confirmation by the British government.[13]

The Governor, however, declined to interfere in the matter on the grounds that the levy was a purely voluntary payment and had never been enforced by previous governments. Langford notes that twelve years later, in 1839, the tax was still being collected on a voluntary basis,[14] but by 1879 it was the general practice for the funds collected from Jain guilds to go to the support of the pinjrapole, while those from Vaishnavas were donated to their temples, from Khojas (Moslem converts of Hindu descent) to the Aga Khan, and from other Moslems to local shrines.

The Gazetteer of the Bombay Presidency describes the pinjrapoles in Ahmedabad District, and the Ahmedabad Pinjrapole itself, at this time. In the Ahmedabad home, for example, the animal population on a day in the beginning of 1875 consisted of 265 cows and bullocks, 130 buffalo, 5 blind cattle, 894 goats, 20 horses, 7 cats, 2 monkeys, 274 fowl, 290 ducks, 2,000 pigeons, 50 parrots, 25 sparrows, 5 kites (hawks), and 33 miscellaneous birds.[15] As to the operating of the pinjrapole:

> Except in special cases, as when at an auction a butcher is bidding for an animal and no one is able to oppose him, it is not the practice to spend the general funds of the asylum in buying animals. In Ahmedabad, where the revenues are large, many animals are saved from slaughter at the weekly Friday fair not only by private charity but by the *panjarapol* clerk who attends for that purpose. All living things are freely received without fee, though a dweller outside the town who sends an animal will generally, if well-to-do, accompany it with a small contribution towards its support. Creatures dangerous to life are not often brought and would perhaps be rejected. Any young born in the home are not usually sold or otherwise disposed of, but no objection is felt to making use of them for the good of the home.
>
> In the Ahmedabad home, all animals are fed within the walls except milch cows which are taken to graze by herdsmen at a monthly charge of 2s [re 1] a head. At the other homes all cattle able to walk are sent out daily to graze. If their number is small they share the common grazing grounds with the other village cattle. In larger places fields are rented or even bought for pasture. Within the walls animals are tended by a staff of servants and if necessary fed with milk. Surplus animals in the Ahmedabad home are drafted to Ranchera [Rancharda], a village in the Gaikwar's district of Kadi devoted to that object by a late Nagar Seth;[16] from Dholera they are sent to Palitana, and each of the smaller institutions has an arrangement by which its surplus can be sent to some larger and wealthier home. The carcasses of the animals which die in the home are sometimes given, sometimes sold to Dheds, who carry them off for the sake of their skins and bury them.

Besides accommodation for four-footed animals and birds every home, except the very smallest, contains at least one *Jīvat Khāna* or insect room. In Ahmedabad, this is filled chiefly by a servant whose business it is, especially in the rainy season when putrid matter is plentiful, to carry a bag around the streets for the collection of maggots and other small vermin. A little grain for their subsistence is thrown into the room and at the end of each year a new room is opened. The old room is closed for ten to twelve years, and after that, as all life is supposed to have ceased, its contents are cleared out and sold as manure.

The general management of the home and the custody of its funds are left to some leading merchant of the Shravak faith who is practically unfettered except by the obligation to consult on any important matter, a few of his co-religionists. He is assisted, in all except the smallest places, by a paid clerk, *gumasta*, who looks after details, keeps the accounts and generally carries out orders. It not infrequently happens that the ordinary revenue is insufficient to meet the expenditure in which cases the *mahajan* is convened and imposes upon itself an extra subscription.[17]

Such was the situation one hundred years ago, and it differs little from that found in Ahmedabad today, for the traditional ties between the pinjrapole, the Jain community, and the mahājans continue unchanged. The nine members of the managing committee of the pinjrapole are of the Jain faith, and they govern with the aid of an advisory board composed of representatives of various mahājans and trade associations, all of which provide financial support for the pinjrapole from their corporate funds.[18]

Some of the close links between the mahājans and the pinjrapole may be seen in the following account of the Maskati Cloth Market Association provided by Śri Hiralal Bhagwati, President of the Association from 1968 to 1974. The Association was founded in 1910 by the merchants of the Maskati Cloth Market in Ahmedabad to protect the interests of the members from other merchants and the mills. Today, there is an entrance fee of Rs. 1,800 (ca. $250) and membership, currently around 1,800, is virtually essential for cloth merchants in both the old and new cloth markets in Ahmedabad, since only association members are allowed to have shops in these markets. Originally, the membership was totally Jain, but now some 40 percent are Vaishnava Hindu.

Three positions on the managing committee of the Ahmedabad Pinjrapole are reserved for the Maskati Association. From its inception, this guild has imposed on its members a levy or cess known as *Dharmada Khoda Dhor Laga* (translating loosely as "tax for helpless animals imposed in the name of religion") and has donated the proceeds to the pinjrapole. Currently, this *lag* takes the form of a tax of 0.0125 percent on all sales made by members and is collected at the end of the financial year. Since the Maskati Market is the largest wholesale cloth market in India, with an annual turnover in sales around Rs. 2,500 million, a considerable sum of money is generated by the lag. A special committee exists to scrutinize accounts, and a penalty of 25 percent is assessed if the lag is found to have been underpaid. Until 1952, the entire lag was donated to the Ahmedabad Pinjrapole, but since that year this has been reduced to 60 percent of the total. The remainder is disbursed to smaller pinjrapoles in the area, to charitable and religious institutions that aid the poor and needy, and is also used to maintain pārabaḍīs.[19]

Just as the ties between the pinjrapole and the vania community have changed little over the last hundred years, so the activities of the institution would appear to have remained essentially the same as those described in the Gazetteer. The pinjrapole was originally located in Jhaveribad, inside the walls of the old city, and here one still finds the head offices, a pārabaḍī, the jīvat khān, and a lane aptly named Pinjrapole Street. There is also a small clearinghouse for animals from the city waiting to be transferred to one of the outlying branches of the pinjrapole where the animals are normally kept. These five centers support the usual assortment of creatures, the sick and maimed as well as the healthy, that characterize the true pinjrapole.

The Ahmedabad Pinjrapole houses mainly cattle, sheep, and goats. Table 2 shows the numbers of these animals handled by the institution during 1973. The figures for this period are abnormally high, directly reflecting the drought and famine conditions prevailing over much of the western part of Gujarat at the time. The monsoon rains had failed for two years in a row, and agriculture was hard hit. Extreme shortages of water and food were being experienced throughout much of the state, and many farmers were having great difficulty sustaining themselves and their families, let alone their livestock. At times the state government was forced to truck in food and water for the human population. Under these conditions, despite attempts at cattle relief, it was inevitable that livestock should suffer. Rather than see them die, many farmers and peasants who were unable to support their animals placed them in relief camps, goshalas, and pinjrapoles. This situation is clearly reflected in the intake of domestic herd animals by the Ahmedabad Pinjrapole in 1973, when some 22,683 cattle, sheep, and goats were added to the existing population of 4,086—a considerable total by any standards!

Under normal conditions, according to the pinjrapole management, herd animals placed in the institution are largely useless—too old, sick, or otherwise disabled to be of much economic value. But in 1973 much of the intake was directly due to the drought, and the animals were left in the pinjrapole not because of any serious physical disability but because of the acute scarcity of fodder. Many owners, however, placed their animals in the asylum as a last resort, and by this time they were suffering so

TABLE 2

AHMEDABAD PINJRAPOLE SOCIETY
ANIMAL REGISTER, 1973

Center	Last year's balance				This year's intake				This year's births			
	Adult cattle	Young cattle	Adult goats, sheep	Young goats, sheep	Adult cattle	Young cattle	Adult goats, sheep	Young goats, sheep	Adult cattle	Young cattle	Adult goats, sheep	Young goats, sheep
1. Ahmedabad	6	3	21	–	33	21	527	192	–	–	–	–
2. Vastrapur	416	183	13	60	2,101	2,023	4,854	2,443	–	–	–	–
3. Rancharda	1,462	377	1,086	227	1,367	4,541	1,084	3,496	–	50	–	21
4. Nesdi	95	97	–	–	–	–	–	–	–	5	–	–
5. Mulsana	9	–	–	–	1	–	–	–	–	–	–	–
6. Vansajda	25	6	–	–	–	–	–	–	–	4	–	–
7. Total	2,013	666	1,120	287	3,502	6,585	6,465	6,131	–	59	–	21

TABLE 2–Continued

Center	Received from other centers				Total			
	Adult cattle	Young cattle	Adult goats, sheep	Young goats, sheep	Adult cattle	Young cattle	Adult goats, sheep	Young goats, sheep
1. Ahmedabad	–	–	–	–	39	24	548	192
2. Vastrapur	69	47	544	190	2,586	2,253	5,411	2,693
3. Rancharda	1,798	1,627	4,673	2,257	4,627	6,595	6,843	6,001
4. Nesdi	111	519	–	–	206	621	–	–
5. Mulsana	257	3	–	–	267	3	–	–
6. Vansajda	2	1	–	–	27	11	–	–
7. Total	2,237	2,197	5,217	2,447	7,752	9,507	12,802	8,886

Center	This year's deaths				Return of boarded animals				Transfers to other centers			
	Adult cattle	Young cattle	Adult goats, sheep	Young goats, sheep	Adult cattle	Young cattle	Adult goats, sheep	Young goats, sheep	Adult cattle	Young cattle	Adult goats, sheep	Young goats, sheep
1. Ahmedabad	–	–	–	–	–	–	–	–	37	20	544	190
2. Vastrapur	754	575	917	427	17	29	–	–	1,518	1,410	4,473	2,259
3. Rancharda	3,243	5,910	5,810	5,791	26	–	1	–	390	325	200	–
4. Nesdi	43	80	–	–	–	–	–	–	62	426	–	–
5. Mulsana	1	–	–	–	–	–	–	–	256	2	–	–
6. Vansajda	–	–	–	–	–	–	–	–	5	4	–	–
7. Total	4,041	6,565	6,727	6,218	43	29	1	–	2,268	2,187	5,217	2,449

TABLE 2—Continued

Center	Total disposal of animals				Remaining balance			
	Adult cattle	Young cattle	Adult goats, sheep	Young goats, sheep	Adult cattle	Young cattle	Adult goats, sheep	Young goats, sheep
1. Ahmedabad	37	20	544	190	2	4	4	2
2. Vastrapur	2,289	2,014	5,390	2,686	297	239	21	7
3. Rancharda	3,659	6,235	6,011	5,791	968	360	832	210
4. Nesdi	105	506	–	–	101	115	–	–
5. Mulsana	257	2	–	–	10	1	–	–
6. Vansajda	5	4	–	–	22	7	–	–
7. Total	6,352	8,781	11,945	8,667	1,400	726	857	219

(Source: Admedabad Pinjrapole Society, *Annual Report, 1973*)

severely from the effects of malnutrition and starvation that their mortality rate was extremely high. Of the 26,769 cattle, buffalo, sheep, and goats in the pinjrapole, some 23,551 (87.98 percent) died during the year.

Despite this loss through natural causes, it can be appreciated that the cost of maintaining these animals was considerable. The Ahmedabad Pinjrapole's accounts (tables 3 and 4) show that expenditures for the year in question amounted to Rs. 1,624,132 (approximately $216,000 at 1973 rates of exchange). These funds were obtained in various ways, but the largest single source of income was the state government, which provided Rs. 592,727 (ca. $79,000) for the maintenance of cattle during the drought period. This was not a regular source of income but was received under the State of Gujarat's cattle relief program, by which a grant of Rs. 1 per day per head of cattle was made to goshalas, pinjrapoles, and other institutions and societies engaged in cattle relief work in the state.

Although the government subsidy was a temporary measure relating specifically to the drought conditions prevailing in Gujarat, the Ahmedabad Pinjrapole continues to receive a regular income from the merchant community, the traditional source of support noted in early accounts of the institution. During 1973, gifts and charitable donations from the public totaled Rs. 316,433 (ca. $42,000), or 19.48 percent of income from all sources,[20] and two-thirds of this sum was derived from mahājan lags. Contributions from specific mahājans were as follows:

Ahmedabad Maskati Cloth Market Association	Rs. 147,627-26
Ahmedabad Seeds Merchants Association	1,002-00
Astodia New Rangoti Cloth Mahajan	751-00
Pañcakua Cloth Mahajan, Ahmedabad	28,201-75
Ahmedabad Grain Merchants Association	9,705-00
New Madhupura Mahajan	501-00
Sugar Bazaar Mahajan	75-00
Old Madhupura Mahajan	551-00
Fatasa Pol Cloth Mahajan	1,043-31
New Silk Cloth Mahajan	6,000-00
Rani's Hajra Cloth Mahajan	3,000-00
Total	Rs.198,457-32

(Source: Ahmedabad Pinjrapole Society, *Annual Report, 1973*)

It is perhaps significant that, while the traditional Ahmedabad textile industry is well represented on the list, all the other mahājans are involved with the marketing of agricultural products.

In addition to the lag contribution by the mahājans of Ahmedabad, the pinjrapole received Rs. 91,954 ($12,225) in the form of general *dharmada*,

TABLE 3

Ahmedabad Pinjrapole Society
Income for the Year Ending 31 December 1973

1972 Rs. Ps.	Income	1973 Rs Ps.	1973 Rs. Ps.
	Rent:		
53,358–17	Rooms	55,735–07	
197,597–24	Buildings	231,372–74	
250,955–41			
2,694–00	Less: Rent refund	244–00	
248,261–41			286,863–81
	Interest:		
	Personal loans	3,102–50	
1,167–73	Securities		
12–67	Bank savings account	28–88	
1,180–40			3,131–38
7,307–56	Dividends:		8,395–96
	Grant:		
	Government animal subsidy		592,726–96
	Donations and gifts:		
185,731–72	Donations (lag) from mahājans	198,457–32	
59,260–28	Dharmada	91,954–00	
49,791–11	Donations for feeding animals	26,022–04	
294,783–11			316,433–36
	Other Income:		
1,935–80	Sale of wool	525–00	
960–97	Sale of ghī	1,314–61	
525–00	Sale of wood	3,506–00	
3,871–06	Fees for boarded animals	2,310–91	
3,768–00	Sale of empty chests and wire	7,932–20	
14,463–50	Sale of dung	15,052–00	
3,056–00	Sale of milk	2,670–00	
114–97	Sale of saffron	393–00	
	Sale of cattle forage	25–00	

TABLE 3—Continued

1972 s. Ps.	Income	1973 Rs. Ps.	Rs. Ps.
	Handling of animals from the municipality	1,035–00	
2,458–00	Income tax refund		
485–00	Sale of iron cart		
31,638–30			34,763–72
	Income from agriculture:		
877–00	Farm association	786–00	
156,583–36	Rancharda income	219,816–20	
96,185–67	Vansajda income	97,758–20	
6,063–00	New Pinjrapole income	13,375–53	
18,094–00	Kasindra income	6,581–05	
11,881–49	Thaltej income	16,599–75	
25–00	Sale of forage		
289,709–52			354,916–73
	Yearly income over expenditure:		
284,913–06	Loan account	25,418–61	
9,960–03	Less: for saving animals		
	Plus: for saving animals	1,481–72	
274,953–03			26,900–33
147,833–33	TOTAL		1,624,132–25

urce: Ahmedabad Pinjrapole Society, *Annual Report, 1973*)

haritable contributions made in the name of dharma, from a variety of ssociations, businesses, and private individuals in the city and else-here. The Ahmedabad Śarafs (Bankers) Mahajan of Manekchowk, for xample, did not collect a lag from its members but instead contributed s. 251 from its funds as a *bhet* or gift. Dharmada was received from Jain aste associations, such as the Jain Swetambara Murti Pujak Poravad aste Society and the Ahmedabad Jain Swetambara Murti Pujak Visa rimali Friendly Society, while the Jain Visa Oswals' contributions came s *lagra prasang,* offerings made at the time of weddings.[21] The greater art of these donations originated in the immediate vicinity of Ahmed-

TABLE 4

AHMEDABAD PINJRAPOLE SOCIETY
EXPENDITURES FOR THE YEAR ENDING 31 DECEMBER 1973

1972 Rs. Ps.	Expenditures	1973 Rs. Ps.	1973 Rs. Ps.
	Expenditures related to property:		
10,006–34	Municipal property tax	9,555–61	
10,454–79	Land tax	5,597–65	
21,584–38	Repairs	37,008–31	
8,686–78	Salaries	9,407–61	
6,898–89	Electricity	6,942–80	
57,631–18			68,511–9
	Expenditures related to management:		
49,772–44	Management salaries	61,930–32	
4,967–70	Travel	6,795–48	
1,032–10	Telegrams and postage	680–55	
4,143–01	Stationery	4,513–36	
1,961–45	Telephone	2,024–90	
9,638–13	Gratuity	2,381–97	
1,880–95	Vehicles	2,049–90	
7,785–68	Provident fund	11,519–60	
	Administrative expenses relating to provident fund	216–85	
16,680–51	Drivers' salaries	19,247–38	
97,861–97			111,360–3
5,330–93	Court expenses		3,577–0
1,300–00	Audit fee		1,000–0
2,041–13	Subscriptions and fees		934–8
15,839–20	Miscellaneous expenses		19,727–9
1,873–90	Daily interest		5,075–4
4,588–00	Depreciation		4,588–0
	Agricultural expenditures:		
146,756–08	At Rancharda	174,803–71	
108,120–75	At Vastrapur	128,552–48	
2,677–05	At New Pinjrapole	1,787–22	
18,219–41	At Thaltej Vid	25,928–45	
23,200–05	At Kasindra	18,424–31	
298,973–34			349,496–1

TABLE 4—Continued

1972			1973			
Rs.	Ps.	Expenditures	Rs.	Ps.	Rs.	Ps.
		Expenditures related to trust purposes:				
		Care of animals:				
202,431	96	Purchase of grain	423,269	90		
112,016	88	Workers' salaries	133,424	01		
					556,693	91
15,468	65	Medicines	14,858	67		
207,839	68	Purchase of hay	162,747	61		
80,684	03	Purchase of dry forage	50,711	68		
3,249	39	Miscellaneous expenses	3,922	78		
4,029	51	Purchase of milk for young goats	5,215	26		
6,522	33	Purchase of water	5,275	26		
12,165	04	Purchase of sugar cane	202,479	69		
1,676	73	Feeding of dogs	1,672	91		
12,537	15	Saving of animals from slaughter	23,402	40		
658,621	35				470,286	26
3,772	33	Additional expenses related to dairy	85,124	96		
		Less: Total dairy income	52,244	69	32,880	27
,147,833	33	TOTAL			1,624,132	25

ource: Ahmedabad Pinjrapole Society, *Annual Report 1973*)

bad, but some were received from as far away as Indore, Bombay, and ven Madras.

Contributions to the pinjrapole often commemorate specific occa-ions. The New Madhupura Mahajan, for example, gave Rs. 100 at the ime of two festivals, Bakr Īd and *Makarśankrant.* Bakr Īd, more com-nonly known throughout the world of Islam as *Id al-Kabir,* is a Moslem estival. Celebrated on the tenth day of Dhu'l-Hadjdja in the Moslem alendar, this is the Feast of Sacrifice when it is the duty of all Moslems vho can afford it to sacrifice animals, usually camels, sheep, and goats, ut sometimes also cattle. The practice originated in the old Arab custom of sacrificing on this day in the valley of Minā (pilgrimage to Mecca, 'Arafāt and Minā is the fifth of the five pillars of Islam), but it has come to be accepted by the Islamic world at large. This slaughter of animals, however, offends the religious sensibilities of devout Hindus, Jains, and Parsees, especially in Gujarat and western India, where ahiṃsā is held in high respect. Some of these communities have even established "Bakr Īd Funds" to be used for the purchase of animals destined to be sacrificed.[22] The New Madhupura Mahajan's donation to the Ahmedabad Pinjrapole at Bakr Īd, therefore, is in the nature of a gesture against Moslem sacrifice at the time of this festival in keeping with the vania community's traditional distaste for the taking of life. Such gestures, moreover, are not limited to the occasion of Bakr Īd. During 1973 a sum of Rs. 20,771 was accumulated by the pinjrapole for purchasing animals, mainly cattle, from butchers to prevent their being slaughtered.

Makarśankrant, on the other hand, is one of the chief seasonal festivals of the Hindu calendar. Falling in the middle of January, it marks the beginning of the Hindu New Year and is celebrated throughout the country. It is a time of pilgrimage, of ritual bathing in the sacred rivers of the north, and of the flying of kites. Although Makarśankrant is celebrated by both Hindus and Jains, for the latter it is a day on which they fulfill one of their Four Fundamental Duties, that of charity, by giving food and clothing to the poor and by providing fodder to cattle.

Donations received by the pinjrapole may be specifically designated for the feeding of animals. During 1973 the Ahmedabad office of the pinjrapole took in Rs. 26,022 for this purpose, while receipts from the Vastrapur and Rancharda branches increased this sum to Rs. 55,150. The Saraspur Sthanakvasi Jain Society of Ahmedabad, for instance, gave Rs. 46-40 for the feeding of dogs. The Seth Jesinghbhai Kalidas Trust of Calcutta donated Rs. 55, of which Rs. 40 were to be spent on fodder for cattle and the remainder on grain for pigeons in the pārabaḍī. The Pipardi Pol Jain Monastery (*upashray*)[23] was one of several such institutions making similar contributions in the name of jīv-dayā.

Lag, dharmada, and donations for the feeding of animals represent the main sources of income derived from public charity. There exists, however, what might be called the *tithi* system, a feature that seems to be exclusively Jain in origin, that adds to the cash income of the pinjrapole. The word *tithi* means "day," and under this system Jains assume certain financial responsibilities for a given day. Money is deposited in a tithi fund; the capital is not touched, but the interest on the capital is used for the feeding of animals in the pinjrapole on a day designated by the donor, usually to commemorate the birthday or death anniversary (*samvatsari*) of some close relative. The tithi fund of the Ahmedabad Pinjrapole amounted to Rs. 358,881 ($50,920) at the beginning of 1973, and during the year four additional contributions totaling Rs. 13,003 were received. The interest on one of these donations, a sum of Rs. 1,001, was to provide food for dogs on the third day of the dark fortnight of Bhadra, the eleventh month of the Hindu year, in perpetuity.

Public donations for the support of animals in the pinjrapole is not limited to money alone. During 1973, various quantities of hay (*bhūsā*), millet (*bājrā*), barley (*jowār*), wheat (*gehūa*), maize (*makaī*), and gram (*chana*)—all itemized in the annual report—were received by the institution for the feeding of its animal population.

As noted earlier, only some 20 percent of the Ahmedabad Pinjrapole's total income during 1973 was accounted for by charitable donations and gifts from the public. Yet, indirectly, the public is responsible for a much greater proportion, since charitable contributions in the past have included not only money but property and land generating a sizable annual income. It is the custom for wealthy vanias, especially if they have no heirs, to will buildings and land to the pinjrapole. Over the years, the institution has acquired considerable holdings. Rent from buildings and rooms donated to or bought by the pinjrapole realized 17.66 percent of the institution's annual income for 1973.

A further 21.85 percent was accounted for by agricultural income derived from the 4,860 acres of land belonging to the pinjrapole. This land, acquired from several sources over a period of years, is scattered around twenty villages in the Ahmedabad region, the main centers being at Rancharda and Vansajda. The land is either farmed by the institution itself or is rented out to tenants both on a cash and a sharecropping basis. Some 1,300 acres at Mulsana are held on a ninety-nine-year lease, but the rest is owned outright, being either purchased out of pinjrapole funds or donated to the institution by various patrons.[24] The pinjrapole's own landholdings are exempted from the Agricultural Tenancy Act, whereby tenants who cultivate rented land for a period of three consecutive years are entitled to its ownership.

In addition to land and property, the pinjrapole's assets include investments in stocks and shares and in savings accounts yielding interest and dividends used to cover operating costs. Given the substantial nature of some of the donations to the institution, the value of its assets, accumulated over a period of years, is not inconsiderable—in 1973 they were worth Rs. 2,687,429 (well over $350,000).

One pertinent feature of the Ahmedabad Pinjrapole's financial situation, in the light of the debate over the utilitarian versus the religious nature of pinjrapoles and goshalas in India, is the minimal contribution that the animals themselves make toward the upkeep of the institution. If the agricultural activities of the pinjrapole are excluded, income received directly from animals and animal products during 1973 amounted to a mere 1.9 percent of the actual cost of supporting the animals in the pinjrapole. This income was received largely from the sale of dung, milk, ghī, and wool. Normally animals are not sold by the institution, though occasionally male calves are put up for auction. Even the carcasses are not sold but are removed by Chamars, the untouchable leatherworking caste, who use the hides to make sandals and other goods.

Some of the pinjrapole's activities have economic aspects, but these are relatively insignificant in terms of the institution's overall operations. One of these is the *khuraki* system run by the pinjrapole. Owners of cattle who, because of shortage of facilities or similar factors, cannot keep their cattle at home may lodge them at the pinjrapole on the payment of boarding fees, which vary between three to five rupees a day, depending on the animal. The users of the khuraki system, usually private individuals who maintain house cows, retain ownership of their animals and can reclaim them at any time. During 1973, thirty-three animals were being kept under the khuraki system—mainly milk cows but also some sick animals and one kid goat.

The pinjrapole also operates a small dairy of some forty-six cows, all bred in the institution itself. The daily output is around 140 liters of milk, which is sold through a coupon system at a rate of Rs. 1-90 per liter. The 350 workers at the institution may purchase milk at the reduced rate of Rs. 0-50 per liter. In 1973 the dairy's income came to Rs. 52,245 ($6,966); dairy expenses, however, totaled Rs. 85,125 ($11,350), leaving a deficit of over $4,000.

The operational expenses of the Ahmedabad Pinjrapole during 1973 (table 4) are of little import for the purposes of this discussion, since they merely reflect the economics of running the institution rather than the social and cultural factors behind it. Suffice it to say that the largest single item of expenditure was the care of animals. The sum spent on workers' salaries, medicines, fodder, water, and the purchasing of animals to save them from slaughter came to over one million rupees. Additional expenses relating to property, administrative costs, and agricultural activities raised the total operating costs to Rs. 1,624,132 or $216,550. This is indeed a considerable budget for an institution that is afforded only brief mention in the existing literature as being of "marginal economic significance."

It should be remembered, however, that the Ahmedabad Pinjrapole Society is not the typical pinjrapole of India. With assets of over two-and-a-half million rupees, an annual income of over one-and-a-half million rupees, and the care of over 20,000 animals, the institution is the largest of its kind in the country. With its roots in medieval times, with its location in one of India's wealthiest cities, whose inhabitants are commonly regarded as best representing the typical Gujarati, and with the scale of its operations, the Ahmedabad Pinjrapole has afforded the opportunity for a detailed discussion of the working of the pinjrapole. It remains to be seen to what extent, in this respect at least, the Ahmedabad institution is typical of other pinjrapoles in India.

The Rajkot Mahajan Pinjrapole

The Ahmedabad Pinjrapole gives a unique view of an old institution in an urban setting in which some traditional features of medieval mercantile communities have survived. This situation is different even from that of the other major urban centers of India, such as Bombay and Calcutta, since the pinjrapoles there have lost much of their original character. The more typical location for the institution is the small town and city of Gujarat, and it is one of these which is examined in the following pages.

To the north and west of Ahmedabad, the nature of the countryside changes rapidly. Fertile fields give way to the arid reaches of the Thar Desert and to the low-lying, seasonally inundated mudflats of the Rann of Kutch. These areas are characterized by low population densities, infrequent settlements, and a nomadic population of camel, sheep, and cattle herders. To the southwest lies Kathiawar, a great peninsula jutting out into the Arabian Sea and bound by the marine embayments of the Gulf of Cambay and the Gulf of Kutch. Deccan lavas, intersected by swarms of trap dykes and flanked by younger sedimentary formations, dominate the region, which presents an aspect of undulating plains and tablelands. Though most of Kathiawar lies below 180 meters, such hills as the Gir Range (640 m.) diversify the landscape, and the highest point in the peninsula, Girnar, reaches 1,117 meters. Rainfall in Kathiawar varies from 400 mm. along the coast to over 1,000 mm. near Girnar. Average precipitation is barely adequate for agriculture, given the high temperatures commonly experienced in the region. The location of Kathiawar relative to the mechanisms of the Indian monsoon, moreover, means that rainfall is often highly variable, especially in the interior districts.[25]

Kathiawar's peripheral geographic location has not isolated the region from the main currents of Indian history. Indeed, the peninsula's westerly location has placed it in the forefront of contacts with southwest Asia from earliest times. The area came under the influence of the Indus Valley culture, and the impressive docking facility at Lothal (ca. 2,000

B.C.) attests to a thriving maritime trade.[26] It was along these same coastal trade routes, almost three millennia later, that Arab traders first brought the religion of Allah to the shores of India. These peaceful contacts, however, were merely harbingers of things to come, for the eleventh and twelfth centuries saw the incursions first of Mahmud of Ghazni and later of Mahmud of Ghuri into western India, incursions that led ultimately to the establishing of Moslem political control over the region.

Despite several centuries of Moslem suzerainty followed by British paramountcy, Kathiawar remained a conglomeration of small, fragmented Hindu states until the middle of the twentieth century. One of these small states, Rajkot, provides the setting for the second case study of pinjrapoles in India. With the incorporation of the princely states into the Union of India after Independence, Rajkot State became part of Saurashtra and then, after the states' reorganization of 1956, of Gujarat State. The town of Rajkot, capital of the former state and headquarters of the British Western India States Agency, became the District Town of Rajkot District and today has emerged as a major regional center with a population over 300,000.[27] In addition to its administrative functions, the city is an important rail junction, has commercial air service, and has a thriving commerce based on cotton, grains, oilseeds, hides, fabrics, and ghī. Its manufacturing industries include flour milling, wool carding, chemicals, matches, soaps, perfumes, and metalworking.

A hundred years ago, however, when the mahājans of Rajkot founded the Rajkot Mahajan Pinjrapole, the city was overshadowed by nearby Jamnagar and the religious center of Junagadh. With a population under 30,000,[28] Rajkot resembled many of the small towns still found throughout the Kathiawar peninsula. Like these towns, Rajkot contained an influential Jain minority, and it was this community that established the pinjrapole in 1880 specifically for the purposes of jīv-dayā. That the original aims of the institution are still being served may be seen in the variety of animals housed in the pinjrapole today.[29]

As at Ahmedabad, many of these animals are bought with funds specifically set aside to save them from being slaughtered. Of the animals in the institution in March 1975, for example, some 150 bullocks and buffalo, all sick, lame, or otherwise of little economic value, had been purchased to prevent their slaughter. Medical treatment is provided by the resident veterinarian, a retired government officer, and the animals are maintained at the pinjrapole until they die from natural causes.

The great majority of the animals found in the pinjrapole, however, are left by the general public. In many instances, the reason for this appears to be the need to avoid personal responsibility for the death of the creature itself—that is, to avoid violating ahiṃsā. During one morning, while I was visiting the pinjrapole, three kid goats were placed in the institution. One of the donors, a Rajput by caste from Kangesiari, a village two miles from Rajkot, explained that he had no use for the goat, a two-day-old male. He would neither kill the animal nor sell it to the local butchers (invariably Moslem by religion), so he brought it to the pinjrapole. No entrance fee was required, though it was customary for some donation to be made according to one's means, and he deposited eight annas (ca. 6¢) in the *dharmada peti,* a box for dharmada contributions located at the gateway to the pinjrapole. Significantly, although the Rajput would not eat beef, he would eat goat, fish, or chicken if it were bought in the bazaar. Hence, at a time when the monsoon rains had failed for three consecutive years and the region was in the grip of a severe drought and famine, he was ignoring a food source, the income from a potential sale, and was even donating cash for the sake of a religious principle.

Ironically, however, the donor was probably indirectly responsible for the animal's demise. Of the 101 goats in the pinjrapole, 75 were kids, most of them brought in when they were only a few days old. Even though they were fed a daily ration of 500 grams of milk, mortality among them was in the region of five to eight animals a day. Despite these losses, the number of goats in the institution was remaining relatively stable, indicating a steady intake of the animals into the institution. It is also interesting to note that most of the goats in the asylum were males, a measure of their low economic potential and standing among the local populace.

During the same morning, another type of donation was received by the pinjrapole. A young woman and her brother, Mehtas of the Jain Dasa Śrimali caste, came as they did every year to commemorate the anniversary of the death of another brother by making an offering in the name of jīv-dayā. They had purchased Rs. 20 worth of green fodder and chana in the local market and had brought it to the pinjrapole to feed the cattle and the pigeons.

As at Ahmedabad, these contributions in cash and kind represent a major source of income for the Rajkot Pinjrapole and are indicative of the support the institution receives from the community at large. Table 5 shows that during the Samvat year 2,014 (A.D. 1958), under relatively normal conditions unaffected by drought, charitable donations accounted for 45.05 percent of the pinjrapole's annual income. Several kinds of dharmada were received. As well as a general dharmada contributed by the public at large, a *mandvi dharmada* was collected from shopkeepers,

TABLE 5

RAJKOT MAHAJAN PINJRAPOLE
INCOME FOR 1958

Income	Rs. Ps.	Rs. Ps.
Current maintenance fund account:		
General dharmada		1,862–00
Dasa Śrimali Mahajan—ulat donations	4,633–00	
—utar donations	1,151–00	
		5,784–00
Rajkot mandvi dharmada	1,190–00	
Donations for the purchase of grass	15,603–00	
Donations for the purchase of oilcake	4,940–00	
Donations for providing milk to goats	137–00	
Donations for feeding corn to pigeons	452–00	
Donations for feeding dogs	141–00	
		22,463–00
Rajkot Pinjrapole trust fund: Interest	9,988—00	
Less: loan interest	1,462–00	
		8,526–00
Bank interest	17–00	
Income from donation box	69–00	
Income from sale of grass	8,515–00	
Income from contract for removing dung	1,025–00	
Income from sale of carcasses	5,472–00	
Income from sale of animals	1,414–00	
Rent of animals for agricultural purposes	929–00	
Income from sale of dung	549–00	
Miscellaneous income	646–00	
Stud fees	40–00	
		18,676–00
Rent from pinjrapole property:		
Palitana travelers lodge and warehouse	2,316–00	
Bhanulal Ramajbhai block building	96–00	
Parekh Bhudarbhai room	18–00	
Pinjrapole building	295–00	
Ratanbhai Bhagwanaj building	54–00	
Harakhbai Punjabhai building	273–00	
Sangharaj block rooms	161–00	

TABLE 5—Continued

Income	Rs. Ps.	Rs. Ps.
Old Pinjrapole	365—00	
New Pinjrapole	1,400—00	
Diwanpura building	754—00	
Amritlal Virpal block	238—00	
New Lane rooms	916—00	
		6,886—00
Rent of land		165—00
Chibhada pinjrapole income:		
Sale of hides	144—00	
Sale of animals	382—00	
Rent of animals for agricultural purposes	6—00	
Sale of dung	436—00	
Donations	426—00	
Rent of land	556—00	
Miscellaneous	64—00	
Donations for the purchase of fodder	515—00	
Sale of grass	140—00	
		2,669—00
Additional income		2,046—00
TOTAL		69,077—00

(Source: Rajkot Mahajan Pinjrapole, *Annual Report, 1958*)

mainly Jains, in various Rajkot markets. A third type of dharmada was donated in the name of members of the Dasa Śrimali Mahajan. Whereas in Ahmedabad most of the mahājans supporting the pinjrapole were trade associations, this particular mahājan was a caste association of Dasa Śrimali Jains.

The donations from the Dasa Śrimali Mahajan fall into two categories—*ulat* and *utar*. Ulat may be interpreted literally as "joy from within" stemming from some happy occasion; in that moment of joy, in order to celebrate the event, announcements are made of contributions to various charities. Ulat donations may be associated with any celebration but are most commonly made at the time of marriages by the family of the bridegroom. Contributions may be made to temples, orphanages, educational establishments, and other charities, but the pinjrapole has traditionally ranked high on the list. Utar donations, on the other hand, refer specifically to charitable contributions made by the bride's family on the occasion of a wedding and, once again, the pinjrapole figures prominently on the list. Both ulat and utar donations are announced publicly by the heads of the respective families to a gathering of male relatives, usually after the actual wedding ceremony and just prior to the departure of the bride and groom for the latter's home. The words *ulat* and *utar* would seem to be local terminology, since Jains from other parts of India and even Gujarat have professed no knowledge of them, but the actual practice is common in Jain marriages throughout the country.

It is perhaps a subtle comment on the relative status of male and female in India that ulat contributions to the Rajkot Pinjrapole are itemized in full in its report while utar contributions are listed only in totals. But even taken alone, ulat donations provide insight not only into some spatial dimensions of money flow associated with the pinjrapole but also into the marriage field of Rajkot. Map 8 shows the frequency and sources of ulat offerings to the pinjrapole over the period of 1952–1958. As might be expected, the highest frequencies are to be found in the nearby towns such as Gondal, Jamnagar, Jetpur, and Wankaner, though contributions are received from all over the Kathiawar peninsula as well as from Ahmedabad, Bombay, and even Calcutta.

In addition to the various kinds of dharmada mentioned above, the pinjrapole also received Rs. 20,543 specifically for the purchase of grass and oilcake for cattle, and lesser amounts to provide milk for young goats, grain for the asylum's two pārabaḍīs, and food for dogs.

As in the case of the Ahmedabad Pinjrapole, charitable donations to the Rajkot Pinjrapole have included land and property, which together generated 22.50 percent of the income for 1958. At present, the institution owns twenty-four acres of agricultural land at Rajkot and another seven acres at its branch at Chibhada, a village some twenty miles west of the town. All this land is devoted to raising barley and millet for fodder. In 1958 the sale of surplus grown on this land realized Rs. 8,655, but in 1975, with the large intake of animals due to the drought and lack of irrigation facilities, the pinjrapole was forced to turn to the open market for its fodder supplies. Under normal conditions this land is used primarily for the support of the institution's cattle. Although its agricultural income is subject to the vagaries of climate, a steady income is provided by the rent on 112 rooms and shops in buildings owned by the pinjrapole, a sum that at present amounts to ca. Rs. 25,000 annually.

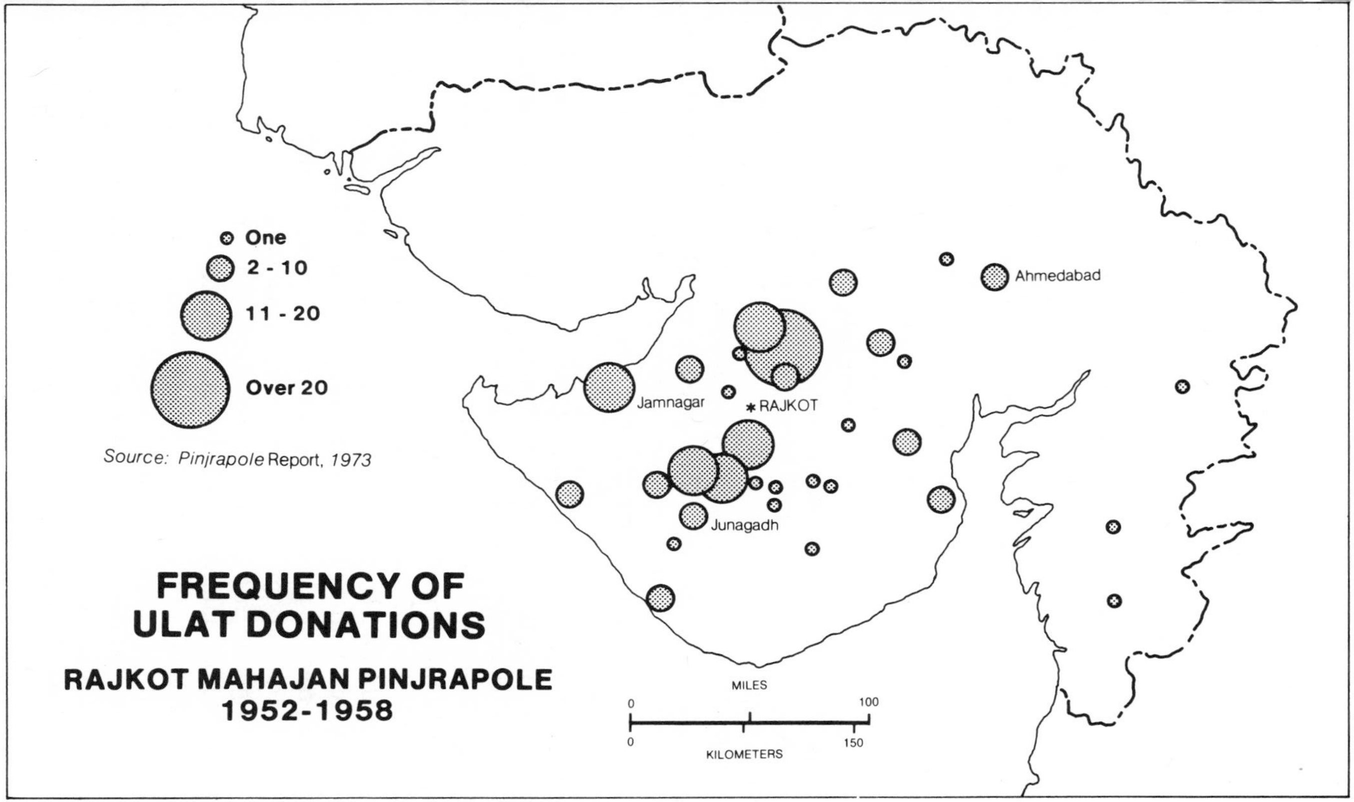

Map 8. *Frequency of Ulat Donations to the Rajkot Mahajan Pinjrapole, 1952–1958*

Over the years, the Rajkot Pinjrapole has also accumulated a sizable trust fund. In 1958 this was worth Rs. 371,896 and earned Rs. 9,988 as interest on permanent deposits, to which might be added interest on bank deposits as well as on loans made from pinjrapole funds to private individuals, usually members of the Pinjrapole Society.

Unlike the Ahmedabad institution, the Rajkot Pinjrapole does not have a milk herd. Nonetheless, despite the fact that most of the cattle in the home are of little economic value, they do contribute indirectly to the upkeep of the institution. Income is derived from the sale of dung (Rs. 985) and from a contract for the removal of dung from its land (Rs. 1,025). Animals in good condition are sold to the public, and the working bullocks of the pinjrapole are rented out to farmers for agricultural purposes. The bulls in the institution are, for a small fee, made available for the servicing of cows brought in by villagers from the surrounding area. When cattle die, they are sold to local Chamars for Rs. 25 per carcass. The Chamars are responsible for the removal of the dead animal and will either use the hide themselves for the manufacture of leather goods, or will sell the carcass to a tannery. Income from these sources totaled Rs. 10,397 or 15.05 percent during 1958.

The operating expenses for the pinjrapole for 1958 are set out in table 6. Feeding costs represented 51.51 percent of total expenditures, and utilities, administrative costs, the renting of pasture, salaries for the 28 paid workers, and building maintenance made up the bulk of the remaining expenses.

In size and location the Rajkot Mahajan Pinjrapole is far more prototypical than the Ahmedabad Pinjrapole. Yet the two institutions share many common features: both are organized by the Jain vania community; both receive financial support from mahājans; and both are maintained in part by charitable donations made by the Jain and, to a lesser extent, the Vaishnava communities in the name of jīv-dayā. Since many of the animals kept at both institutions are noneconomic in nature, the pinjrapoles would seem to be fulfilling what is essentially a religious function in protecting life according to the Jain credo of ahiṃsā. Even the cattle in the institutions have little economic potential beyond the value of their dung and hides. The preservation of their lives, with many actually being bought to save them from the slaughterhouse, would seem to be an expression of ahiṃsā rather than of any particular feeling for the cow or regard for its sanctity, even though other communities, such as the Hindu, that support the institutions may be motivated by such considerations. Certain features suggest that religious associations between the cow and Krishna are not completely absent. The Rajkot Pinjrapole

TABLE 6

RAJKOT MAHAJAN PINJRAPOLE
EXPENSES FOR 1958

Expenses	Rs. Ps.	Rs. Ps.
Current maintenance expenses:		
Purchase of green grass	17,859–00	
Purchase of grass	958–00	
Purchase of oilcake	6,148–00	
Purchase of milk for goats	5,056–00	
Feeding of pigeons	1,516–00	
Maintenance of dogs	635–00	
Purchase of dung	234–00	
Transportation expenses	266–00	
Medicines	870–00	
Expenses of letters of congratulation	8–00	
Printing charges	56–00	
Telegrams and postage	9–00	
Telephone	146–00	
Advertising	146–00	
Bonuses	278–00	
Building repairs	3,566–00	
Workers salaries	5,639–00	
Allowances	38–00	
Electricity	1,696–00	
Lothari Bhayasar camp	38–00	
Rent of pasture at Lothari Bhayasar	14,835–00	
Water costs	291–00	
Dharmada costs	28–00	
Miscellaneous costs	1,138–00	61,454–00
Chibhada pinjrapole expenses:		
Bonus	20–00	
Advertising	15–00	
Workers salaries	1,953–00	
Allowances	48–00	
Dung	269–00	
Sheep	622–00	
Cultivation costs	1,023–00	
Miscellaneous	76–00	

TABLE 6—Continued

Expenses	Rs. Ps.	Rs. Ps.
Oilcake	762–00	
Grass	2,477–00	
Maintenance of goats	4–00	
Building costs	354–00	
		7,623–00
TOTAL		69,077–00

(Source: Rajkot Mahajan Pinjrapole, *Annual Report*, 1958)

has a shrine to Krishna, and the Gopāṣṭamī festival is celebrated by go-pūjā and a procession of cows in the pinjrapole—not unexpected in a part of the country with strong historical ties to Krishna. Yet both the Ahmedabad and the Rajkot pinjrapoles remain essentially Jain institutions, run and supported by the Jain community for purposes stemming from the traditional Jaina concern for the sanctity of animal life.

VI

The Gift of a Cow—Case Studies of Goshalas in Modern India

Cows are God, they seem to me to be Indra.
Cows are the draught of the first–pressed soma.
These very cows, O men, are Indra.
I long for Indra with my heart and thought.
Ṛg Veda 6.28.

For the Jain, the cow is nothing more nor less than *jīva,* a life form requiring the general respect accorded to all life forms. But for the Hindu, the cow is special, something akin to the gods and to be treated as such. Indeed, reverence for the cow and cow worship represent persistent motifs pervading much of Hindu culture and society. This chapter examines some of the manifestations of Hindu attitudes toward the cow by presenting detailed studies of three goshalas in India today.[1]

Śri Nathji Goshala, Nathdwara

After the halcyon days of Akbar's reign, the peoples of India were forced to suffer the depredations of Aurangzeb, that scourge of Hinduism, who ascended the Peacock Throne in 1659. Conceiving it his duty to put an end to the religious tolerance that had characterized the rule of Akbar and Jehangir, Aurangzeb set out to destroy Hinduism, ordering that all schools and temples of the infidel be demolished and that their teachings and religious practices be put down. In order to escape this persecution, the devotees of Krishna from Mathura, the capital of Vraj, fled their city, taking with them one of the most revered of their *mūrtis* or images, that of Śri Nāthji (fig. 11). Reputedly dating from the time of Krishna's deification between the eleventh and twelfth centuries before Christ, the Śri Nāthji image was supposedly set up in a shrine by King Vajra of the Yādavas when he was brought to Mathura by Arjuna, the hero of the *Mahābhārata* epic, after the destruction of the Yādava race. The statue remained underground in Goverdhan Mountain for many centuries until it was discovered and placed in a temple built on top of the mountain by Vallabhāchārya. To escape Aurangzeb's wrath, the descendants and followers of Vallabhāchārya left Vraj with their image and, after wandering through Rajasthan, were invited to Mewar by that state's ruler, Rāna Rāj Singh. Tradition has it that on the journey to the capital at Udaipur, the wheel of the chariot carrying the mūrti of Śri Nāthji sank deep into the earth and defied all efforts to extract it. This incident was seen as an omen from the god, Krishna, that he wished to reside at that spot, and so a temple was built to house his image. The nearby village of Siarh and its land were given in perpetual gift by the Rāna of Mewar to the Mahārāj Gosain, the spiritual head of the Vallabhāchārya sect of Vaishnavas.[2]

Around the temple has grown up the town of Nathdwara, "The Portal of the God," described by the Rajputana Gazetteer as:

> This famous shrine of Krishna . . . situated 22 miles north-east of Oodeypore [Udaipur]; it is a large walled city on the right bank of the Banas: on the north-east and south it is surrounded by hills, but to the west across the river, which here takes a sharp bend, it is more open. . . . large herds of cattle graze on these hills to the east of the city, where is a regular cattle-farm surrounded by a high wall and guarded by sepoys.[3]

The cattle farm is, in fact, the Śri Nathji Goshala, the temple goshala of the Śri Nāthji shrine at Nathdwara.

As temple goshalas are attached to the temples themselves, historical records are not readily accessible and documented evidence concerning age is difficult to come by. The Nathdwara temple, for example, is relatively young by Indian standards, being founded in 1671, and the goshàla was supposedly established at the same time. But officials of the temple argue that the temple goshala is at least 400 years old, reasoning that since all Vallabhāchārya temples have goshalas, there must have been one when the Śri Nāthji shrine was first set up near Mathura.

Documentary evidence for this is not available, but other sources from the sixteenth century suggest that this might have been the case. The following is the text of a *farman*, or document of royal grant, issued by

Fig. 11. *Śri Nāthji (Krishna), from a Rajasthani Temple Hanging (after Skelton).*

Akbar's Commander-in-Chief in favor of the temple of Goverdhan Nāth before the shrine was removed to Rajasthan:

GOD IS GREAT

Order of Khan Khannan, Disciple.
Khan Bahadur, Commander-in-Chief.

> Be it known to the present and future Officers of the Paraganah, that as in the villages of Savi, etc., there is the grazing land for cows and oxen belonging to Gordhan [the Temple of Goverdhan Nath], they should not prohibit and instruct them on the ground of watching charges and counting the head of cattle, because the villages have been purposely given in grant. They should act in conformity with the Order of the Exalted [One], and take action accordingly. On no pretext should a new permit be demanded every year. Written on Roz Azar, Azar month Elahi 33rd [Regnal] year, corresponding to the 11th of the month of Moharram ul Haram A.H. 997. (1st December [N. S.] A.D. 1588).[4]

Thus, almost four centuries ago the Goverdhan temple owned herds of cattle, presumably for reasons that hold true today, with grazing rights granted by the Moghul Emperors. In an earlier farman dated A.D. 1581, issued by Akbar himself, Vithaleshwa, the second son of Vallabhāchārya, was given the right to graze his herds on all lands in the area, whether they were *khalsa* (hereditary) or *jagir* (granted to an individual in return for services to the royal court). Another farman some ten years later notes that Vitalrai (Vithaleshwa) had purchased land at Jatipura near Gokul and "caused to be built thereon buildings, gardens, cowsheds and Karkhanas (workshops) for the Temple of Gordhan Nath. . . . " and grants him and his descendants freedom from all taxes in perpetuity.[5]

Today the activities of the Nathdwara temple goshala would seem to differ very little from those of its precursor of four hundred years ago. Milk from the goshala, which cannot be used in ways not related to the temple since the institution is regarded as belonging to Krishna himself, is taken to the temple kitchens in large earthenware jars carried on the heads of servants. There it is made into various types of milk products and sweetmeats, which are then offered to the deity as prasād. At Nathdwara, as at all Krishna temples, the daily obeisances take the form of eight darśans, the showing of the god's idol to the public, each representing some aspect of the god's daily activities. Thus, the first darśan is *Mangāla*, the morning levee, which takes place just after sunrise when the god (in the form of his idol) is awakened and bathed in milk. The second, *Śṛiṅgāra*, occurs one-and-a-half hours later, when the god is dressed in his jewels and robe and seated on his throne. At these and at each succeeding darśan, devotees gather at the temple to see and worship their god. Before each darśan, prasād is offered to the deity and afterward, having been blessed by the god's acceptance, it is distributed in lieu of salary to the *sevites* of the Nathdwara temple, the people who perform services in the temple and the temple kitchens. These temple servants often sell their prasād to the many pilgrims who visit the shrine from all over the country.

The goshala also fulfills a ceremonial role in the religious life of the Nathdwara temple since it is the scene of the Gopāṣṭamī festival, sometimes known as *Gocharan*. Gocharan literally means "the grazing of the cows" and commemorates the occasion when the boy, Krishna, first took his foster father's cattle to graze in the forests of Vrindavan. Because the festival falls on the eighth day of the bright fortnight of the month of Kartik (September–October), it is more commonly referred to as Gopāṣṭamī (*aṣṭamī*=eighth).

At the Śri Nathji Goshala, the animals are washed, decorated, and fed special foods such as *duliya* (mash) and *gur* (molasses), often donated by the public for the occasion. Pilgrims to Nathdwara and villagers from the surrounding districts come to the goshala for darśan of the cows, bringing with them *luddus* and other delicacies to feed the animals. The head of the temple comes to the goshala in procession and performs gopūjā, the cow-worship ceremony. And, perhaps most interesting of all, the goshala stages bullfights in honor of Krishna (figs. 12 and 13).[6] Dignitaries and special guests are seated on a circular stone platform in the center of the large inner courtyard of the goshala, the gates are closed and two bulls are led in. Each is branded on both flanks, with the circle of the Sun-god, Sūrya, on one side and the *triśul*, or trident, of Shiva on the other. Cheered by the onlookers, their gentle eyes belying the impression of power in their imposing humps and massive bodies, they butt each other and lock horns in a test of strength until one turns tail and flees, pursued by the other. The two bulls are then separated by cowherds of the goshala, no mean feat in itself, as the animals are not permitted to harm each other. The goshala maintains ten zebu bulls specifically for the purpose of fighting and breeding.[7]

Other Krishna-related festivals are celebrated at the goshala, though in a less elaborate manner. *Annakut*, or Goverdhan Pūjā, occurs a week before Gopāṣṭamī, on Kartik Sud 1. This commemorates the time when Krishna saved his people by lifting Goverdhan Pahar (mountain) on his little finger and sheltering his followers and their herds from the rains of Indra. An image of Krishna is fashioned from cow dung (*gobar*) and is worshipped by the devotees who have come to celebrate the festival. Then a herd of cows with peacock feathers—a symbol of Krishna—tied to their horns is driven across the image to destroy it. At the same time, bullfights are staged at the goshala. A third festival, Janamaṣṭamī or

Fig. 12. *Bullfight Mural at Śri Nathji Goshala, Nathdwara.*

Fig. 13. *Bullfight at Śri Nathji Goshala, Nathdwara*

Krishna's birthday, is also celebrated at the goshala though this tends to be kept largely by the cowherds who work in the institution.

The activities of the Śri Nathji Goshala clearly revolve around the worship of Krishna, both in religious festivals observed at the institution and also in its provision of milk for the Krishna shrine at Nathdwara. As it is attached to the Śri Nāthji temple, it is managed and financed through the temple itself, which comes under the auspices of the *Devasthan*, a state government organization that oversees the running of temples throughout Rajasthan. The goshala, which is operated at an annual cost of ca. Rs. 500,000, receives its income from the temple, and all charitable donations for the goshala itself go directly into temple funds.

Goshala lands, too, are temple property, and the 5,000 acres are held as *thikana* (religious grant) from the erstwhile state of Mewar. This land is used for grazing and for the production of fodder to feed the goshala's cattle, so the purchasing of fodder is necessary only in times of famine. There are just under 1,000 animals at the Śri Nathji Goshala, most of them at Nathdwara itself and the remainder in branches located elsewhere on temple lands. These animals either have been donated to the goshala by devotees of Krishna or have been bred in the institution itself. Like most temple goshalas, the Nathdwara goshala differs from the vania institution in that it does not normally accept useless animals from the general public. The management reports that some ten years earlier it had been the practice to provide refuge to nonproductive cattle and that the goshala had maintained some 1,500 head of old and disabled animals, but that this had since been discontinued. When goshala cattle grow old, they are not disposed of, but are allowed to die a natural death, so that at present there are approximately forty old or lame animals in the institution.

One atypical feature of the Nathdwara institution is the keeping of buffalo—some 150 animals—specifically for milking purposes. Buffalo milk is valued for its higher fat content, even though it is regarded as ritually inferior to cow's milk. It is kept separate from cow's milk and is used only in the preparation of foodstuffs, not in temple ritual or for the making of prasād—a fine example of the relative status of the buffalo vis-à-vis the cow in India today.

In addition to providing milk, the animals of the goshala contribute to the running of the temple by providing fuel for the temple kitchens. The kitchens prepare prasād for the deity and meals for the many pilgrims that visit Nathdwara from all over the west of India. The dung from the goshala is gathered in a field opposite the goshala gates, where teams of women pat and shape it into cakes that are left in the sun to dry. When the cakes are ready, the women load them into baskets, put the baskets on their heads, and descend the hill toward the town and the temple precincts. These women are usually the wives and relatives of the cowherds who work in the goshala. In total, there are some forty workers who, with their families, are employed in the institution. These servants receive no salary but are paid in kind, being given a daily allowance of milk from the goshala and also a monthly allotment of grain, *dāl* (lentils), salt, chilis, and cloth.

This, then, is the temple goshala of the Śri Nathji temple at Nathdwara, one of the most important of the Krishna shrines in western India. Its prime function is religious in nature, namely, the provision of milk to the temple for use in Vaishnava ritual. In form, origin, and organization it is quite different from the Marwari or vania institution.

Śri Panchayati Goshala, Vrindavan

> A fine country of many pasture-lands and well-natured people, full of ropes for tethering cattle, resonant with the voice of the sputtering churn and flowing with buttermilk: where the soil is ever moist with milky froth, and the stick with its circling cord sputters merrily in the pail as the girls spin it round—in homesteads gladdened by the sputtering churn.

So says the *Harivaṃśa Purāṇa* in a bucolic description of Vraj, that part of the Upper Ganges plains around Mathura in western Uttar Pradesh. A region of considerable cultural and historical importance, Vraj occupied a central location in the states of ancient Madhyadesha, with Mathura, its capital city, being regarded as among the greatest in the land and one of the seven holy cities of India. Frequently mentioned in the ancient and medieval works of India and other countries, the past glory of Mathura is reflected in its former role as a meeting place for foreign and indigenous cultures, as a center of trade and commerce, and in the development of its own tradition in art and sculpture.[8]

Although from earliest times Mathura was an important religious center for Brahmanical Hinduism, Jainism, and Buddhism, these traditions have since been superseded by the worship of Krishna. Today the city is one of the most—if not the most—important centers of the Krishna-*bhakti* cult in the entire country. This prominence arises from the historical associations of the area with Krishna himself. Kaṃsa, the ruler of Mathura, so the legend goes, imprisoned his cousin, Devaki, and her husband, Vāsudeva, after it was prophesied that he would be killed by one of her sons. Six of her children, born in prison, were killed at birth on Kaṃsa's orders, but the seventh, Krishna, was secretly carried across the River Jumna to Gokul on the northern bank, where he grew up as the hangeling son of Nandi Gopa, the cowherd chief. Many a tale is told bout the youthful adventures of Krishna and his brother, Balarāma, mong the cowherds and *gopis* (cowgirls) of Vraj, and hardly any place f significance in the area is not sanctified through its associations with rishna's mischievous escapades and occasional supernatural achieve-ents. Goverdhan Pahar, Gokul, Nandgaon, Radhakund, Mathura itself, e birthplace of Krishna, all bask in the glory of the cowherd god, and one more so than the town of Vrindavan.

Located some ten kilometers upstream from Mathura, Vrindavan emains one of the most important pilgrimage centers of the Krishna cult, r it was in the forests of Vrindavan that Krishna spent many hours nding his father's herds of cattle, playing with his cowherd friends, or allying with the gopis. The town itself is surrounded on three sides by e sacred Jumna River, the river front being covered with a series of *hats,* or steps, leading down to the water's edge. It is on one of these hats, the Kāliya Mardan Ghat, that one finds the *kadam* tree (*Nauclea adamba*) from which, according to local tradition, Krishna jumped into he river to combat the serpent, Kāliya; further downstream at Keśi Ghat s the location of another legendary exploit, the killing of the equine onster, Keśi.

Today, Vrindavan has a resident population of 25,000, but this umber is continually swelled by the influx of devotees from all parts of he country who come to visit the scene of Krishna's childhood and some f the more than 1,000 temples and shrines that are found within the own's limits. At all times of the year, and especially on the occasion of mportant festivals, the road from Mathura is lined with pilgrims making heir way into the town on foot, by *tonga* (horsedrawn carriage), or by rightly painted bus. Many, en route, pause to worship or donate money n the small shrines found in the several goshalas they pass on their ourney, and it is one of these which provides the second case study in his chapter.

The Śri Panchayati Goshala in Vrindavan is a vania goshala although, s its name implies, it was founded in 1850 by the town Panchayat *pañcāyat)* or local council. According to the members of the current anagement, the goshala was founded in the name of dharma; the cow vas a sacred animal and the especial favorite of Krishna, thus it was the luty of every true Hindu to look to its welfare. It was, moreover, fitting hat such considerations of dharma should be of prime concern in /rindavan with its unique associations with Krishna, and to this end the anchayat, all Krishna followers, had decided to set up the goshala over century ago.

As in the case of the Nathdwara goshala, the links with the cult of

Krishna are clearly reflected in the emphasis placed on Krishna-relate festivals, which figure prominently in the religious calendar of th region. Some, such as Janamaṣṭamī and Goverdhan Pūjā, are widel celebrated in villages and temples throughout the area as well as i goshalas, but in the Vrindavan institution—as at Nathdwara—Gopāṣṭan is the most important event of the year.

At the time of Gopāṣṭamī, the goshala is decorated for the festivitie (fig. 14). Two small palm trees are set up on either side of the gat streamers of the sacred *tulsi* leaves (basil), garlands of flowers, an multihued buntings are hung across the entrance and veranda, and large banner proclaiming the occasion is suspended across the street. I the morning a procession of cattle is taken through the streets of the city Some one hundred of the goshala's best cows are selected, washed and decorated for the parade. Their horns are painted silver, their bodie are daubed with orange dots or handprints,[9] and some have bunting and garlands strung around their necks. The herd is led by a zebu bul with flowers twined around his horns and neck.

As in legend, Krishna and his brother, Balarāma, led their herds out t graze in the forests of Vraj, so in fact their cattle are taken through th streets and alleyways of Vrindavan by two *swarups,* or incarnation of Krishna and Balarāma. The young boys who play these roles ar dressed in the manner of their illustrious predecessors, with long flowing, sequined robes and golden crowns decorated with peacock feather fans. Their foreheads are marked with the sign of Krishna an their faces are covered with sequined patterns. Around their arms ar silver bracelets, and on their feet they wear silver anklets covered wit tinkling bells. Krishna carries with him his everpresent flute.

Led by a group of drummers accompanied by an escort of "cowherds" bearing the ubiquitous orange banner of Krishna, the herd sets of toward the center of the town. In the rear come the swarups and a crow of devotees, many of them women from far off Manipur State who hav made the pilgrimage to Vrindavan. The procession winds its wa through the narrow streets of the city, completely disrupting traffic an forcing passersby to take shelter in doorways, while people pause fro their tasks to watch the cows go by or even to offer them food. Onlooker kneel in the dust and bow their foreheads to the ground, or they touc their fingers to the dung, mute evidence of the passing of the herd, an place a smear on their forehead. Why? The cow is the home of all th gods, and thus the very dust through which it walks is holy. Moreover these are Bhagwan Krishna's cows that he is taking out into the woods t graze—the dust and dung are doubly holy.

After wending its sinuous way through the alleys and bazaars o

Fig. 14. *Gopāṣṭamī at Śri Panchayati Goshala, Vrindavan.*

Vrindavan, the procession returns to the goshala's premises, where th festivities continue. In the courtyard of the goshala a shrine has been s up under a canopy. The centerpiece, placed on a low table, is a frame colored photograph, hung with a *mālā* of marigolds, of a painting Krishna performing gopūjā before a cow. In front of the picture is smaller table covered with a red cloth on which is placed a bowl flowers and tulsi leaves. In the middle of the bowl lies a hollow, ova shaped piece of wood (*golā*) containing a stone (*nārāyaṇ*) represen ing the god, Ganesh, the remover of all obstacles. The bowl is placed the center of a series of designs on the table traced out of rice. These tak the form of a square, divided into quarters with the bowl containing th nārāyaṇ at the center. In three of the quarters are placed small piles c rice, each having symbolic meaning, and in the fourth, that ancier Aryan sign, the swastika, is also made out of rice.[10]

Worship is performed before the swarups of Krishna and Balarām who are then seated on the shrine on either side of the picture of Krishn to oversee the rest of the ceremony (fig. 15). The officiating priest, with tray containing a red powder (*rang*), flowers, betel nut (*pān*), *perās* (sweetmeat made from milk), incense, tulsi leaves, lengths of red an yellow string, and a ghī lamp, then offers pūjā at the altar while intonin Sanskrit mantras. Mālās are placed around the nārāyaṇ and offerings c perās, pān, and betel leaves are placed on the shrine. The priest, using cluster of tulsi leaves for the purpose, sprinkles the altar first wit pañcamṛit and then with water from the sacred Jumna River. Followin this, the colored strings are placed on the altar and the people come t offer their own prasād—fruit, sweetmeats, pān, betel leaves, and money.

The ceremonies at the shrine completed, the proceedings now move t a nearby tree, for here, waiting patiently with her young male calf, is th cow selected for gopūjā. Interestingly enough, the goshala cow chose for this honor is not a pure Indian breed but is a Jersey cross.[11] Gopūjā i performed not by the Brahman who has been leading the worshippers a the shrine but by the guest of honor who, on this occasion, happens to b a wealthy Marwari vania from the city. In the same manner as Krishna i the picture, the vania offers gopūjā to the cow, which is sprinkled wit holy water, garlanded, and painted with a *tikka* or red dot on its forehea (fig. 16). Then the merchant and his entourage walk around the cow sev en times in a clockwise direction, pausing to touch her forehead an pressing the tip of her tail to their own foreheads in passing. This cir cumambulation, or *parikramā*, is completed by crawling under the bell of the cow.

Gopūjā completed, the scene shifts back to the shrine, where devotees continue to make offerings to Krishna, placing tikkas on their own

Fig. 15. Swarups *of Krishna and Balarāma.*

foreheads and tying the colored threads around their wrists, the left wrist for men and the right wrist for women. The ceremony is concluded with the performing of ārtī before the shrine, accompanied by the chanting of appropriate mantras. The morning's activities are brought to an end by the distribution of pañcamṛit to those who have attended the celebration; a small amount of the liquid is poured into the devotee's right hand and is drunk, with the wet hand then being wiped over the head.

The Gopāṣṭamī festivities adjourn until evening, when perhaps eight or nine hundred people come to the Panchayati Goshala for darśan of the cow. They belong to all levels of the community, from wealthy merchants and businessmen to peasants and villagers, from city residents to pilgrims from all over India. With them they bring special foods such as *chappatis* (bread), luddus, and gur to feed to the cattle. They light ghī lamps before the cows, performing their own personal ārtī and gopūjā, and they place sticks of burning incense in the cow dung that lies scattered over the courtyard of the goshala. But the biggest attraction of the evening is the *Krishna Līlā*, a folk drama recreating those events of Krishna's life commemorated at the time of Gopāṣṭamī. The Krishna Līlā is an ancient tradition of the Mathura region, reputedly derived from the sporting of Krishna with the gopis, and today there are several professional companies that stage performances at various times of the year. An impromptu stage is erected in the goshala court and elaborately costumed actors, many of them children, reenact for an enthralled audience the events of Gocharan. At some institutions, cows are decorated and led to the stage at the appropriate moment for the performance of gopūjā (fig. 17).

Whereas Gopāṣṭamī is usually celebrated only in goshalas, the second major Krishna festival to be held at the Vrindavan goshala, Goverdhan Pūjā, is also enacted in Krishna temples. The following account describes the ceremony as witnessed in the Radha Raman Mandir (Temple) in Vrindavan, though it is virtually identical to those held at goshalas throughout the Mathura region. The centerpiece of the ceremony is a larger-than-life-sized image of Krishna, his scarf flowing, his legs crossed and his left hand upraised, supporting Goverdhan Mountain—all fashioned out of cow dung (fig. 18). The image, laid out on the floor, is elaborately decorated with rice, colored powder, turmeric, and flour. A stone from Goverdhan Mountain is placed in Krishna's navel and the whole image is enclosed in a rectangular frame of dung. At the four corners are placed straws with clumps of cotton attached to their tops to represent the forest. On the walls of the room are hung *pichhavaī*, large tapestries, often worked with gold and silver thread, depicting the lifting of the mountain and other episodes from Krishna's life (fig. 19).

Fig. 16. *Cow worship (Gopūjā)*

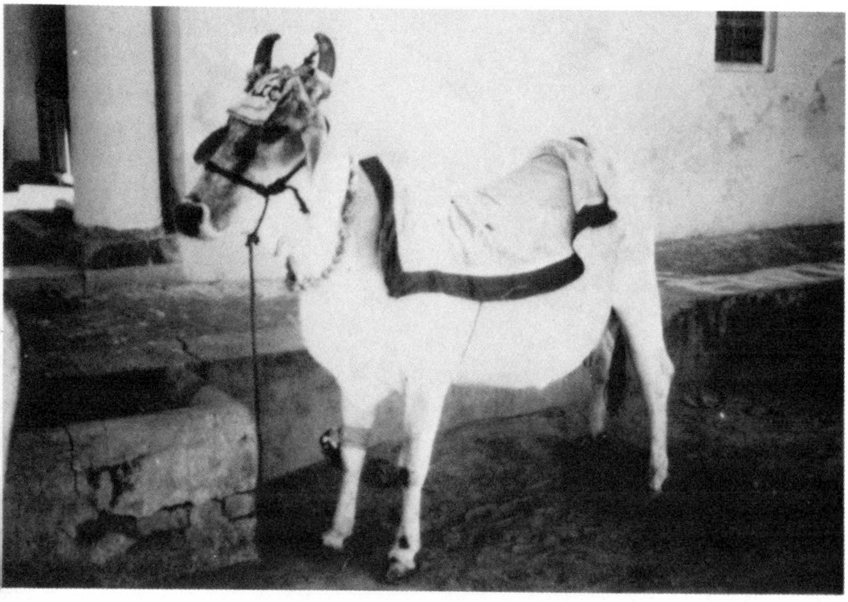

Fig. 17. *Goshala Cow, Vrindavan*

Fig. 18. *Dung Image of Goverdhan Nāth (Krishna).*

Fig. 19. *Temple Hanging for Goverdhan Pūjā.*

The first part of the ceremony in the temple, something not found ir the goshala, involves the appearance of the image of Krishna on a silve throne accompanied by a host of animals—elephants, deer, bull—al crafted out of silver. Then the officiating priest and his assistants, al Brahman followers of Chaitanya, proceed with the invocation to Krishna A small calf is already in the temple waiting to play its part in th ceremony, and, after the invocation, the priests perform pūjā to the cal and then to the dung image of the deity. While chanting mantras, milk i poured over the stone in the image's navel, ghī lamps are lit and place on the image, and the priests perform parikramā—walking around an around the image in a symbolic circumambulation of Goverdhan Moun tain itself, a devotional task that many Krishna followers actually per form on this day, sometimes even by repeatedly stretching themselve out on the ground until they complete the entire circuit on thei stomachs.

After the parikramā by the priests, the remainder of the participant place various offerings of perās, luddus, flowers, and money on th image, ghī lamps are lit, and the audience performs parikramā to th strains of devotional music played by a small group of musicians.

The ceremony ends when the calf is driven across the dung image i order to destroy it. Once the invocation to Krishna is made, so th reasoning for this goes, the image is god incarnate and it would be sinfu for mere man to destroy it. The calf, however, as the animal beloved o Krishna, can do so with impunity; once it has walked over the image, th image loses its divine identity and the dung can be removed. Finall pañcamṛit is offered to the devotees attending the festival, and th abundance of prasād is distributed among the priests and their helper

As at Nathdwara, Gopāṣṭamī and Goverdhan Pūjā are two importan festivals illustrating the strong ties between the goshala and the worshi of Krishna. These celebrations, however, represent only one aspec albeit a very colorful one, of this persistent association; another may b seen in the format of the goshala's management and membership.

Today the Panchayati Goshala is managed by the Panchayati Goshal Committee, of which one becomes a general member by contributing R 1,100 to the goshala. Presently there are some two hundred gener members and a working committee of thirty-three who handle the routin affairs of the institution. The great majority of the members are Vais navas, followers of Vishnu in his incarnation as Krishna. In addition, th membership shows distinct community affiliations. In 1973, for exampl both the president, the late Śri Biharilal Jhunjhunuwala of Vrindavan, an the secretary, Śri Nand Kishore Jhajharia of Calcutta, were Marwa vanias whose families originated in Rajasthan.[12] The remaining men ers of the committee, whether resident in Vrindavan or Calcutta, were verwhelmingly Marwari vania by community. Only seven of the thirty-three man committee, Brahmans by caste, did not belong to the Vaiśya astes. Of the remaining twenty-six, two were what might be termed Agarwals of local origin, and the remainder were Marwaris, Agarwals oming originally from the Marwar region of Rajasthan. All of the ommittee were Krishna followers.

The composition of the 1973 managing committee reflects the essen-ial nature of the Vrindavan goshala—a vania institution supported by he Hindu community in general but particularly by Marwari business astes, in which the worship of Krishna is well entrenched. These ommunities, moreover, provide the financial support for the goshala hrough their donations. Unlike the Ahmedabad and Rajkot Pinjrapoles, o formal lag or cess is collected in the bazaars. This is a measure, erhaps, of the dominance of the religious functions of the city over the ommercial. Donations to the goshala, accounting for 43.66 percent of ncome in 1972, take the form of khuraki (merely contributions made for he maintenance of cows, and not the system of boarding cattle found at ne Ahmedabad Pinjrapole), general contributions for the assistance of he goshala, and donations made specifically for the purchasing of milk ows. Monthly contributions, whereby donors agree to give a certain sum ach month, provide additional funds, and collection boxes are placed at arious locations throughout the city, mostly in the many *dharamśālas*, ee inns for pilgrims, that abound in this sacred city of Krishna.

The single most important source of funds for the Vrindavan Goshala s Calcutta, with Delhi and Bombay following as poor second and third. ome of the contributions received, especially from the three premier ities of India, are regular donations from various charitable trusts. In 973, for example, the Śri Vrindavan Gosevā Trust of Calcutta made a ontribution of Rs. 10,000 as khuraki to the goshala. But the great ajority of donations are from private individuals and are made at the me of personal visits to the institution. It is significant that most of the onors of khuraki, assistance, and even monthly contributions come not om the Vrindavan area but from the length and breadth of northern ndia, some even from as far as Assam and Karnataka (map 9). This xtensive "generative" field can be explained in terms of the historical ssociations of the Mathura area, including Vrindavan, with Krishna, nd the region's importance as a focus of Vaishnava pilgrimage. The ominance of Calcutta over Bombay and over Delhi, only 100 miles to the orthwest, reflects the concentration of Vaishnava and Marwari commu-ities in that great industrial and commercial city.

If public donations are responsible for a major proportion of the

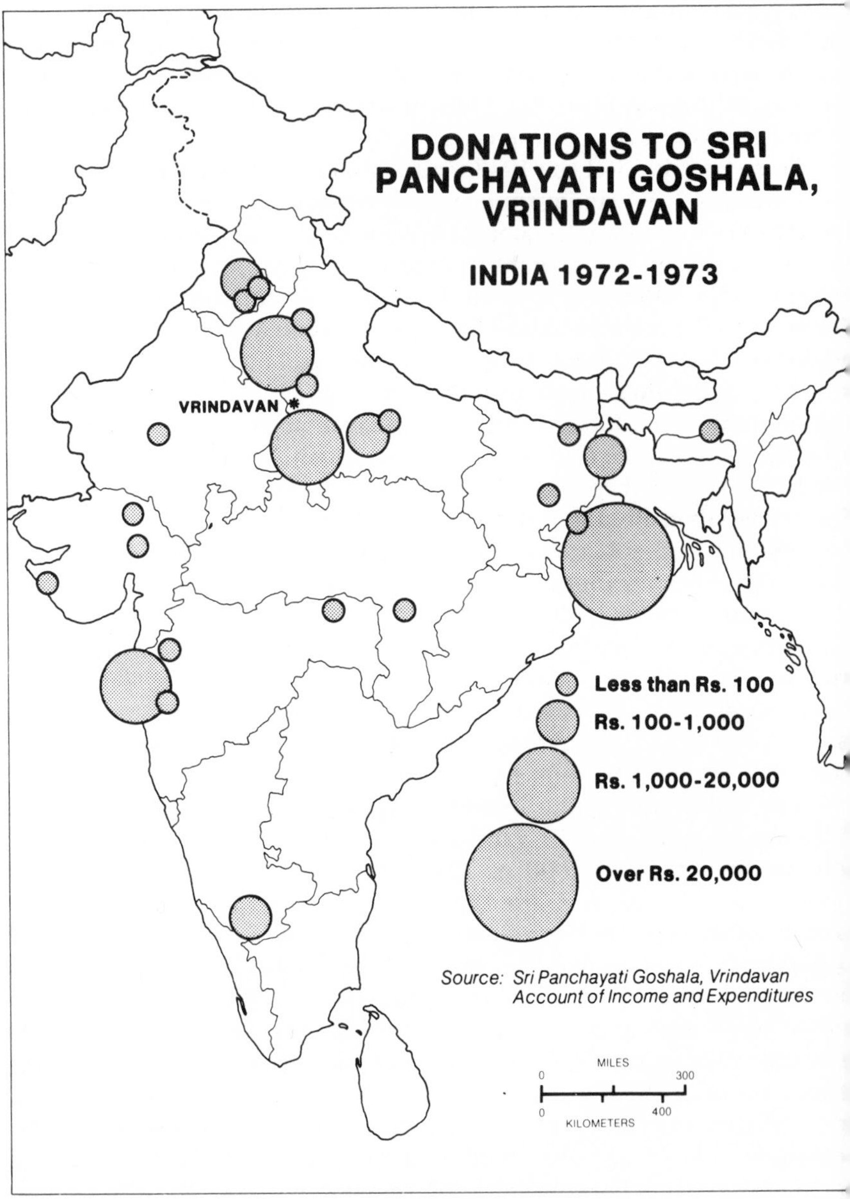

Map 9. *Donations to Śri Panchayati Goshala, Vrindavan, 1972–1973*

Vrindavan Goshala's income, the animals of the institution also constitute an important source of revenue. The goshala is of medium size, having in November 1974 a total of 383 animals. Unlike the Ahmedabad and Rajkot Pinjrapoles, with their concern for a broad spectrum of animal life, and unlike the Nathdwara Goshala, with its milking buffalo, only cattle are kept at the Vrindavan institution. Not all of these, however, are useless, since the goshala maintains a sizable milk herd. Some 62 cows are too old, sick or otherwise disabled to be productive. Of the remainder, 54 cows are in milk and another 48 between lactations or in calf; there are, in addition, 7 stud bulls of Haryana breed, 4 working bullocks, 133 heifers, and 75 calves.

The goshala serves as a dairy as well as an animal refuge—a dual function that appears to be increasingly common in modern times. Several of the cows in the institution have been purchased specifically for milk production, and an additional thirteen of the milk herd have been received as *godān*, ten on the occasion of weddings and three at the time of deaths. The practice of godān, literally "gift of a cow," is a survival of an ancient custom whereby cows are given in payment to Brahmans who officiate at marriages and death ceremonies. Today, the gift of a cow has generally been replaced by cash, but some Hindus continue the custom by donating cows to the goshala.

Most milk animals are either bred in the institution or are recovered from the goshala herd. Salvageable cows left at the goshala because their owners can no longer afford to keep them are served by the goshala's breeding bulls and are added to the dairy section. Of all the calves born in the goshala, only the male calves are sold to farmers and peasants for agricultural and draught purposes; all the females are kept and, when old enough to calve, are added to the milking section. The males, sold at the age of around one year, bring funds into the goshala. In 1972 thirty-two male calves were sold for Rs. 3,092 (table 7).

The sale of milk produced by the Vrindavan Goshala's dairy section accounts for 22.77 percent of the institution's total income for 1972. The milk is sold to the aged, the sick, and to orphanages in the city at the rate of Rs. 2-50 per liter, which is about the market rate, using an advance sales system. Most of the 200 customers pay one month in advance and then come to the goshala every day to collect their milk. A home delivery service is available at an extra charge of Rs. 2 per month, with milkmen employed by the goshala making rounds of the city on bicycles twice a day.

The nonproductive animals of the pinjrapole section of the institution are segregated from the dairy herd. These cows are left in the goshala by their owners, mainly villagers from the surrounding districts but also

TABLE 7

Śri Panchayati Goshala, Vrindavan
Income for the Year Ending 5 November 1972

Income	Rs. Ps.	Rs. Ps
Donations:		96,708–69
Khuraki	56,814–62	
Goshala assistance	28,085–76	
Donations for purchasing milk cows	6,102–00	
Gopāṣṭamī	3,036–16	
Monthly contributions	1,780–00	
City land gift	706–35	
Collection boxes	183–80	
Sale of milk, calves, dung cakes and manure:		59,425–27
Milk	50,663–90	
Male calves	3,092–21	
Dung cakes and manure	5,669–16	
Income from agriculture		12,075–50
Bhima land	7,555–43	
Dhorera land	4,539–05	
Pachistha land	235–02	
Kaiya land	208–00	
Babari land	150–00	
Less expenses	612–00	
Interest and rent:		20,180–93
Pakor building	7,476–09	
Bombay block	10,500–20	
Interest	1,948–34	
Rent of weighing scales	256–30	
Tractor:		19,768–69
Earnings	40,246–55	
Expenses	20,477–86	
Additional income:		14,362–89
TOTAL		222,521–97

(Source: Śri Pa hayati Goshala, Vrindavan, *Account of Income and Expenditures, 1971–72*)

city residents as well, who cannot afford to keep their animals but who do not wish to be responsible for their destruction. House cows are also placed in the institution, and occasionally strays are accepted as well. Even the minimal residual value of these useless animals is not fully exploited, since when they die their carcasses are either burned or deposited in the Jumna River.[13] Their dung, however, is sold as manure or made into dung cakes and sold as fuel.

In addition to the production of milk for sale on the open market, the Vrindavan Goshala's agricultural activities include the farming of some 500 acres of land. Of this, some 200 acres are under grazing, and this is supplemented by the practice of allowing goshala cattle to graze in the jungle. The remaining 300 acres are under cultivation, mostly with fodder crops grown for the maintenance of the cattle population of the goshala. Fifty acres of this are irrigated by a tube well and pump, so production is not totally at the mercy of the monsoon rains. During 1974, production totaled 1,200 *maunds* (one maund is 37.39 kilograms) of barley and wheat, 2,000 maunds of hay, and 2,000 maunds of green grass. The barley and wheat were sold on the market and the hay and grass were used for feeding goshala cattle, but the institution still had to purchase from the bazaar 8,000 maunds of hay and 1,200 of green grass, as well as concentrates, at a total cost of some Rs. 100,000.

The goshala's agricultural land produces fodder and earns income (Rs. 12,075 in 1972) from the sale of crops. This income is supplemented considerably by the earnings of the goshala tractor, which is rented out to local farmers during the ploughing and harvesting seasons, once work on goshala land is complete.[14] In addition to the tractor, the agricultural equipment of the goshala includes two ox carts, an electric chaff cutter, a thresher, a harrow and a seed drill.

Besides owning land, the Vrindavan Goshala possesses buildings that are rented out to the public. A dharamśāla in Vrindavan is let rent free, but one house in Calcutta and some property in Bombay generated 8.08 percent of the goshala's income for 1972. To this may be added interest on loans made from goshala funds (Rs. 1,948).

The total income for the Vrindavan Goshala from all sources during 1972 amounted to Rs. 222,522. In addition, the total assets of the institution—land, property, capital, and livestock—was valued in the same year at Rs. 738,614, though according to the management this figure was closer to Rs. 1,000,000 in 1974.

Table 8 shows the operating expenses of the goshala for 1972. The major item of expenditure is animal welfare, accounting for over half (53.89 percent) of the total outlay. The cost of running the dairy section accounts for another 24.12 percent (Rs. 53,682) of the year's expenditure.

TABLE 8

Sri Panchayati Goshala, Vrindavan
Expenses for the Year Ending 5 November 1972

Expenses	Rs. Ps.	Rs. Ps.
Animal welfare		119,935–87
Cow khuraki	99,989–41	
Wages	16,051–59	
Medicines	62–75	
Miscellaneous	3,832–12	
Milk production:		53,631–13
Cow maintenance	41,869–89	
Wages	5,329–35	
Purchasing of cows	6,299–20	
Miscellaneous	132–69	
Administrative expenses:		17,963–96
Wages	11,846–30	
Bank commission	158–90	
Electricity	600–70	
Postage	252–55	
Office supplies	402–90	
Building repairs	1,751–93	
Entertainment expenses	93–73	
Water	418–15	
Cycle repairs	427–57	
Transportation costs	464–19	
House and water tax	94–50	
Telephone	55–75	
Miscellaneous	1,396–79	
Other expenses:		9,875–36
Gopāstamī	858–01	
Dharmàda assistance	875–00	
Laxmi Govind Shiva temple	1,040–00	
Audit fees	400–00	
City taxes	2,302–35	
Uncollectable debts	4,400–00	

TABLE 8—Continued

Expenses	Rs. Ps.	Rs. Ps.
Agricultural expenses:		21,116–65
Rent of land	1,064–60	
Legal expenses	3,350–69	
Wages	5,206–70	
Miscellaneous	2,431–04	
Losses from Samvat year 2028	9,063–62	
TOTAL		222,522–97

(Source: Śri Panchayati Goshala, Vrindavan, *Account of Income and Expenditures, 1971–72*)

Significantly, although the dairy has realized a net profit for the year, this sum (Rs. 6,424) is minimal compared to both receipts through charitable donations and to the overall cost of animal welfare in the institution. Various administrative costs, charitable and religious donations by the goshala, and miscellaneous expenses, including a loss from the previous year, complete the expenditures for 1972.

The Panchayati Goshala at Vrindavan differs significantly from the pinjrapoles at Ahmedabad and Rajkot and from the Nathdwara temple goshala. It is essentially a Marwari Vaishnava institution, managed and supported by the Marwari vania community not only from Vrindavan but from all over northern India. Unlike the pinjrapoles, it serves as a refuge only to cattle, though in recent years the scope of its activities has expanded to include the economic. Through the acquisition of land and property, the goshala pursues agricultural activities and assumes the role of landlord. With the development of a commercial dairy herd, and with the sale of male calves and dung, it further involves itself in the local agricultural economy. Yet, in spite of these activities, the goshala continues to provide shelter to nonproductive cattle, and the persistent links with the cult of Krishna suggest that the religious underpinnings of the institution remain as viable as before.

Śri Gopal Goshala, Dibrugarh

If the Vrindavan Panchayati Goshala is unique in the sense that it lies in the very heart of Vraj, the land of Krishna, the Śri Gopal Goshala in Dibrugarh, a town in eastern Assam, provides an example of a vania institution typical of goshalas scattered throughout the rest of India.

Headquarters town of Lakhimpur District, Dibrugarh is located on the south bank of the Brahmaputra River some one thousand kilometers from its mouth in the Bay of Bengal. With a population of 80,348, it is the easternmost town of any size on the alluvial plains of the Brahmaputra, for only eighty kilometers upstream the river emerges from the hills of Arunachal Pradesh. The foothills of the Himalaya rise from the plains a mere thirty kilometers north of the town, while one hundred and twenty kilometers to the east the mountains sweep southward into upper Burma. The international frontier between India and Burma is only one hundred kilometers to the southeast, and to the south lies the hill country of Nagaland, so that the only route of easy access and contact with the rest of India is southwestward along the valley of the Brahmaputra. Dibrugarh is the eastern terminus of the main railway service from lower Assam, West Bengal, and Calcutta, as it was, formerly, for the steamers that used to ply the Brahmaputra.

Located in upper Assam at the eastern extremity of the subcontinent, the very isolation of Dibrugarh combined with the contrasting environmental conditions of the region are factors in my selection of the Śri Gopal Goshala for the final case study of the goshala in India. Here, unlike the drought-prone plains of Gujarat or the slightly more humid upper Ganges Valley, the well-watered alluvial lowlands of the Brahmaputra receive an abundance of moisture. Early and prolonged monsoons combine with local topography to yield over 2,000 mm. of rainfall on the valley floor and even higher amounts on the surrounding hills. The heavy precipitation and surface runoff create a perennial flooding problem during the summer months. This climatic difference is clearly reflected in agricultural activities, for in Assam, rice replaces wheat and millet as the staple food crop, with tea and jute being the major cash crops. The physical and economic setting of the Dibrugarh Goshala is, therefore, significantly different from those of the institutions discussed earlier.

Culturally, too, Dibrugarh offers meaningful contrasts. Unlike Nathdwara and Vrindavan, Dibrugarh's marginal position relative to the Aryan heartland, and the comparatively recent colonization of the Assam valley by Hindus, enables us to examine an institution on the frontier of the Hindu culture area, perhaps leading to clarification of features less easily discernible elsewhere. Moreover, distance and isolation from the major urban and industrial centers of India provide a setting that has presumably experienced minimal change in the traditional attitudes and values that tend to alter rapidly under the impact of urban and industrial expansion.

Today, in addition to its administrative functions, Dibrugarh serves as a major regional center, its trade and commerce being based largely on agricultural products.[15] Yet, at the turn of the century it was a small town with a population of just over 11,000.[16] The Imperial Gazetteer of India notes that at that time the entire wholesale trade of the town was in the hands of Marwari merchants,[17] and it was this community that founded the goshala in Dibrugarh. Like the great majority of goshalas in India today, the Śri Gopal Goshala is less than one hundred years old, having been established in 1904 by two Marwari vanias, one a Jain and the other a Vaishnava, who had left Rajasthan to pursue their business activities in Assam. The avowed reason behind the establishment of the institution was dharma, the desire to protect the cow because of religious sentiment.

Marwari involvement and support for the goshala continues today. Of the 1970 managing committee, for example, eighteen of the twenty-three members belong to the Marwari community, the remainder being Brahmans. This dominance of Marwaris is seen also among the roughly 3,000 members belonging to the Goshala Society, drawn from Dibrugarh itself as well as from such nearby towns as Tinsukia and Dumduma. It is interesting to note that, according to the goshala management, some fifteen or twenty Moslems make regular donations to the institution, being in sympathy with its aims and objectives.

As of 14 May 1974, there were a total of 270 cattle in the Śri Gopal Goshala. Of these, 40 were in milk and another 59 were between lactations, giving a total milk herd of 99 animals. These included both western crosses as well as *deshi* (nondescript breed) cows, since the institution had 3 western-breed stud bulls among its total of 9 breeding bulls. One, a Friesian cross, had been donated by the Department of Animal Husbandry of the Assam State Government as part of a scheme for improving milk production, and the others, an Australian cross and a Jersey cross, had been purchased by the goshala itself. This introduction of western cattle breeds was part of a specific policy pursued by the goshala to improve its milk herd and to increase milk yields. The results of this may be seen in the rise of milk output from 19,223 liters in 1970 to 40,000 liters in 1974, although part of this is attributable to an increase in the size of the milk herd itself. It does, however, indicate a willingness on the behalf of the management to accept foreign breeds yielding higher returns. The milk is sold at Rs. 2 per liter by a coupon system, whereby customers purchase a month's supply of coupons in advance which are then redeemed for milk, either delivered at home by goshala employees or collected at the goshala by the customer himself. The purchase of milk is open to anyone, not just to goshala members, and in 1974 some 110 customers were obtaining their milk from the institution. The sale of milk represented an important source of funds for the goshala and in 1970 accounted for 26.58 percent of total income for the year (table 9).

TABLE 9

ŚRI GOPAL GOSHALA, DIBRUGARH
ANNUAL ACCOUNT OF INCOME FROM 16 OCTOBER 1969 TO 10 OCTOBER 1970

Income	Rs. – Ps.
Last year's balance	15,422–69
Donations during the Gopāṣṭamī Mela	3,218–00
Offerings to the mūrti (statue of Krishna), receipts from women offering *thālī ka dān* (gift trays) and from anonymous donors during the mela	912–27
Sale of green grass during the mela	261–62
Income from sideshows and sports during the mela	644–11
Income from the lottery during the mela	4,733–35
Donations made in the name of Gopāṣṭamī received after the mela	2,669–25
Donations made to commemorate special occasions	7,300–00
Donations made in the name of Brahmāpuri	1,447–00
Donations made at the time of marriages	1,917–50
Donations of rice straw	8,634–71
Donations for the maintenance of cows	391–00
Donations for providing calves with milk	22–00
Donations for providing straw and grass for animals	187–00
Donations for the construction of buildings	5,001–00
Rent from land and property	3,735–00
Income from collection boxes (dibba) in Dibrugarh	12,383–23
Income from collection boxes in the surrounding area	28,919–86
Income from the sale of milk	39,704–00
Rent of chairs	9,751–60
Income from the garden	5,439–50
Government grant	7,500–00
Remuneration for animal displays	150–00
Commission on the sale of bran	365–00
Income from sale of male calves	975–00
Interest on loans	1,314–50
Income from protecting cows	252–00
Sale of empty sacks	1,226–65
Sale of wood	93–00
Auction	97–00
Sale of old furniture	48–00
Rent of bullock cart	5–00
Income from khuraki system	61–00
Income from sale of grain	26–02
TOTAL	164,807–86

(Source: Śri Gopal Goshala, Dibrugarh, *Annual Report, 1971*)

Although some of the milk animals in the goshala were purchased by the institution itself, many were donated by the public. Some appear to have been received in the form of godān; others were given as charitable donations in support of the goshala, or even as a means of disposing of unwanted animals while ensuring that they were not killed. As table 10 shows, 2 of the animals received in this manner in 1970 were cows, 2 were bulls, 10 were female calves, and 14 were male calves. The females were added to the goshala's milk herd, the bulls were used for breeding, and the male calves were sold to the public. All of the donors of these animals were from Dibrugarh.

In addition to the milk herd, the Dibrugarh Goshala maintained 51 cows that were nonproductive—too old, sick, or disabled to be of any economic value. In 1970, for example, a total of 21 useless animals were left in the goshala by inhabitants of Dibrugarh and surrounding villages (table 11). Stray animals are not accepted because of possible complications with newfound owners, but cattle injured in accidents on the streets will be given a home. No fee is expected when an owner places his cattle in the goshala, though donations at this time for the upkeep of the animal are welcomed. The productive and nonproductive herds are segregated, and potentially useful cows left in the institution are covered by goshala bulls and added to the milk herd. This service is also available to the general public for a nominal fee of Rs. 4.

There does exist at Dibrugarh a khuraki system similar to that found at the Ahmedabad Pinjrapole. Owners of animals, usually house cows, who are unable to maintain them for short periods of time may board them at the goshala upon payment of the requisite fees, and may reclaim them at a later date. Little use is made of this service, the income from this source amounting to a mere Rs. 61 during 1970.

The cattle of the Śri Gopal Goshala are kept at the institution's premises at Dibrugarh, where the Goshala Society owns 40 *bighas* (25 acres) of land. This, plus 180 bighas at Bokel, was acquiried as a result of charitable donations from the vania community and forms an important part of the institution's assets. The land at Bokel is rented out, providing a steady income for the goshala, while the Dibrugarh property is used exclusively for the grazing of goshala cattle. Some additional grazing is obtained on government-owned land, where the goshala's cows are allowed to feed free of charge on stubble after crops have been harvested, their dung being left to manure the fields. The goshala also follows the practice, widespread throughout India, of turning their animals out to forage what they can along the wayside.

As no fodder crops are grown on goshala land, fodder has to be purchased on the open market, and this forms a major item of expendi-

TABLE 10

Śri Gopal Goshala, Dibrugarh
Animals Received as Gifts, 1970

Donor's name	Town	Animal	No.	Description
1. Government of Assam, Department of Animal Husbandry	Dibrugarh	Bull	1	Healthy
2. Śri Binajraj Garodiya	Dibrugarh	Male calf	1	Healthy
3. Śri Pandit Durgadatt	Dibrugarh	Female calf	1	Healthy
4. Śri Srilal Agrawal	Dibrugarh	Male calf	1	Healthy
		Female calf	1	Healthy
5. Śri Asu Babu/Aurora Cinema	Dibrugarh	Female calf	1	Healthy
6. Śri Bhagwandas Goyanka	Dibrugarh	Male calf	3	Healthy
		Female calf	1	Healthy
7. Śri Mahesh Prasad Keriya	Dibrugarh	Male calf	1	Healthy
8. Śri Basudev Bhagat	Dibrugarh	Male calf	1	Healthy
		Cow	1	In calf
9. Śri Maniram Hanuman Baksh	Dibrugarh	Female calf	1	Healthy
10. Śri Bhagwandas Jitani	Dibrugarh	Male calf	1	Healthy
11. Śri Ramchandra Verma Vakil	Dibrugarh	Male calf	1	Healthy
12. Śri Maliram Khemka	Dibrugarh	Male calf	1	Healthy
13. Śri Phoenix Motor Company	Dibrugarh	Female calf	3	Healthy
		Male calf	1	Heathy
		Bull	1	Healthy
14. Śri Nathmal Jain	Dibrugarh	Male calf	1	Healthy
15. Śri Motilal Jitani	Dibrugarh	Male calf	1	Healthy
16. Śri Ramkumar Kanoi	Dibrugarh	Female calf	1	Healthy
17. Śri Mohanlal Keriya	Dibrugarh	Cow	1	Healthy
18. Śri Nareshchandra Datt	Dibrugarh	Female calf	1	Healthy
		Male calf	1	Healthy

Total of animals received as gifts

Cows–2 Male calves–14
Female calves–10 Bulls–2

(Source: Śri Gopal Goshala, Dibrugarh, *Annual Report, 1971*)

ture for the institution. In 1970, some Rs. 51,096, representing 38.02 percent of the year's total outlay, was spent on cattle feed, mainly hay, rice straw, and wheat bran. By 1974 this figure had risen to Rs. 125,000 (see table 12).

Some of this is, of course, covered by the sale of milk, dung, and male calves but, during 1970, total income from these sources amounted to only Rs. 40,679—a mere 30.27 percent of the annual expenditure. Indeed,

TABLE 11

ŚRI GOPAL GOSHALA, DIBRUGARH
INTAKE OF OLD, CRIPPLED AND DISEASED ANIMALS, 1970

Donor's name	Town	Animal	No.	Description
1. Śri Bhagwandas Jitani	Dibrugarh	Cow	1	Emaciated, lame
2. Śri Nareshchandra Datt	Dibrugarh	Cow	1	Old
3. Śri Mahavirprasad Kishanlal	Dibrugarh	Cow	1	Barren
4. Śri Bajranglal Beriya	Dibrugarh	Cow	2	Old, barren
5. Śri Dindayal Kejriwal	Dibrugarh	Female calf	1	Rheumatic
6. Śri Assistant Manager, Green Hood Tea Estate	Green Hood	Cow	1	Ailing
7. Śri Buddhakarang Chaikhani	Ghoghrajan	Cow	1	Ailing
8. Śri Shankarlal Sureki	Dibrugarh	Cow	1	Ailing
9. Śri Maniram Hanuman Baksh	Dibrugarh	Cow	1	Rheumatic
10. Śri Durgadatt Haralalka	Dibrugarh	Cow	1	Old, feeble
		Bullock	1	Old, feeble
11. Śri Chief Health Inspector	Dibrugarh	Bullock	1	Old, crippled
12. Śri Nathmal Jugalkishor	Dibrugarh	Female calf	1	Rheumatic
13. Śri Sanvarmal Sahariya	Dibrugarh	Cow	1	Old, feeble
14. Śri Chiranjilal Keriya	Dibrugarh	Male calf	1	Very weak
		Female calf	1	Very weak
15. Śri Mohanlal Khemka	Dibrugarh	Cow	1	Old
		Male calf	1	Weak
16. Śri Phoenix Motor Company	Dibrugarh	Cow	1	Old
17. Śri Mohanlal Keriya	Dibrugarh	Cow	1	Old

Total of old, crippled, and diseased animals

Cows–14	Male calves–2
Female calves–3	Bullocks–2

(Source: Śri Gopal Goshala, Dibrugarh, *Annual Report, 1971*)

table 9 indicates that the mainstay of the goshala's finances is not the milk herd or agricultural activities but rather charitable contributions from the general public, which accounted for over half (52.58 percent) of the year's total income.

Charitable donations to the Dibrugarh Goshala take many forms, but the most important are those received through the *dibba* system. The dibba is a collection box placed by the goshala in shops, hotels, and businesses throughout the district, some 450 in Dibrugarh itself and another 1,500 in nearby towns. The system appears to function as an informal lag, for the owners of the businesses deposit contributions in

TABLE 12

ŚRI GOPAL GOSHALA, DIBRUGARH
ANNUAL ACCOUNT OF EXPENDITURES FROM 16 OCTOBER 1969 TO 10 OCTOBER 1970

Expenses		Rs. Ps.
Gopāṣṭamī mela expenses		5,828–02
Cost of arranging mela sideshows and sports		121–00
Canteen expenses at the mela		812–55
Salaries and wages		24,667–11
Purchase of straw and grass for cattle		14,835–11
Purchase of grain		36,260–48
Cost of building repairs and construction		16,741–61
Purchase of cows		14,727–70
Repairs to collection boxes, purchase of new boxes and padlocks		847–84
Travel, food and accommodation expenses incurred in distributing collection boxes		1,643–33
Office expenses		888–36
Electricity		1,911–27
Printing costs for 2000 copies of the annual report		2,897–78
Telephone		449–95
Municipal tax (including Rs. 1,050 as arrears for the last seven years)		1,285–92
Expenses relating to soil-erosion prevention		609–58
Repairing of chairs		765–30
Purchase of new chairs		1,771–25
Expenses relating to the garden		1,486–16
Disposal of dead animals		147–25
Medical treatment for the animals		1,440–63
Miscellaneous purchases		4,111–18
Legal costs		95–00
Expenses relating to goshala lands at Bokel		57–75
Total expenses		134,402–13
Loans given by the goshala at a monthly interest rate of one percent per month to the following firms:		20,000–00
1. Śri Motilal Pratapchand and Co., Diburgarh	10,000	
2. Śri Dibrugarh India Tea Co., Diburgarh	5,000	
3. Śri Sitaram Buddhakaran Chaukhani and Co., Ghoghrajan	5,000	
	20,000	
Balance		10,405–73
Total		164,807–86

(Source: Śri Gopal Goshala, Dibrugarh, *Annual Report, 1971*)

the boxes, though customers may also donate in the name of dharmada. The goshala sends workers around to various locations to collect the donations and to repair or replace the boxes themselves; receipts are issued for the sum collected, and listings of amounts are printed in the goshala's annual report. During 1970, Rs. 41,303 (27.65 percent of the annual income) was derived from this source.

Over two-thirds of this amount was collected outside Dibrugarh, from the smaller towns in the surrounding districts. Map 10, showing the locations of the goshala collection boxes, provides an indication of the area from which the goshala derives its public support. Naturally, Dibrugarh and the nearby Tinsukia have the greatest concentration of dibbas, but they are found as far as Pasighat and Margherita, up to eighty kilometers away. As might be expected, the collection boxes are located in the larger towns and villages rather than in rural areas, although they are found on many of the tea estates in the district. As the great majority of participants in the dibba system are Marwaris, it is clear that the goshala receives support not only from the business community in Dibrugarh, but from Marwari vanias throughout the areas. The institution does, therefore, appear to assume a social function, acting as a focus for the business community, and particularly for the Marwari community of the entire region.

Another aspect of the broader role that the goshala plays in the social life of the community is seen in the *mela* or fair held to celebrate Gopāṣṭamī. Although not quite as elaborate as the festivities held at Nathdwara or Vrindavan, it is still the main event of the year, as indeed it is in goshalas everywhere in India. Commemorating as it does an event in the life of the god, Krishna, Gopāṣṭamī is essentially a religious festival, but any festival in India soon becomes a social occasion. Lasting two days, its activities revolve around the worship of Krishna and the performance of gopūjā. A mūrti of Krishna is set up, and people visit the goshala for darśan and to worship the cows. Stalls depicting scenes from Krishna's life are constructed, and various sideshows and sports are held for the amusement of the public. The goshala runs a lottery to raise funds for the institution and sells grass to the visitors for feeding the cows. It is customary for patrons to offer donations to the goshala at this time, and receipts for 1970 show contributions from as far away as Calcutta and even Bombay.

Gifts of money, received by the goshala throughout the year, are often made to commemorate special events, such as the successful completion of important business deals, to mark anniversaries, or to celebrate marriages. One category of donations of special note is those received in the name of Brahmāpuri, a local deity. In addition to money, in 1970 one

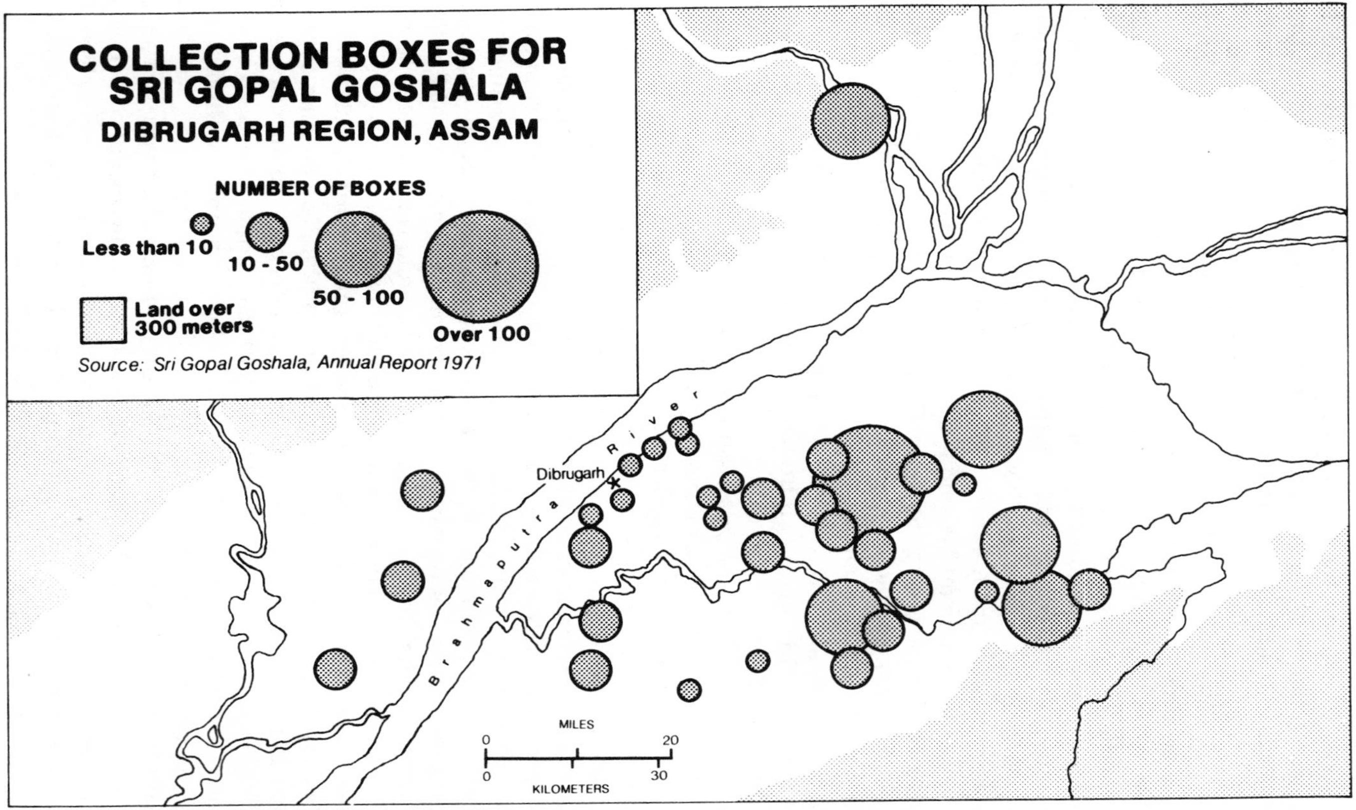

Map 10. *Collection Boxes for Śri Gopal Goshala, Dibrugarh*

Śri Mahavirprasad Kishanlal contributed, in the name of Brahmāpuri, 32 sets of eating utensils—*thālīs* (trays), *kaṭōrīs* (bowls), glasses, and spoons. These were subsequently donated by the goshala to the Marwari Hospital in Dibrugarh. Other contributions were specifically earmarked for providing fodder to cattle, for providing calves with milk, and for construction purposes.

Donations are made in kind as well as in cash. During 1970 the Dibrugarh Goshala received 1,880 kg. of rice straw, 370 bags of raw sugar, 80 kg. of wheat mash, 2 bags of salt, 33 bags of bran, 279 kg. of molasses, 82 bags of chopped straw *(khudi),* and 5,000 bundles of straw. Despite this, however, cattle feed had to be purchased for the animals in the goshala, and the Rs. 51,096 spent on this again represents the largest single item of expenditure on the institution's balance sheet. Salaries, wages, and administrative costs account for the bulk of the remaining operating costs of the goshala (table 12).

A final item of note is the practice of lending money to firms and private individuals out of goshala funds. In 1970, Rs. 20,000 were lent to three businesses—two in Dibrugarh and one in Ghoghrajan—at a monthly interest rate of 1 percent.

In its nature, therefore, the Śri Gopal Goshala in Dibrugarh is similar to the Vrindavan institution, although it lacks the historical associations with the land of Krishna's boyhood. It is multifunctional, being involved in agriculture and milk production as well as in cow protection, but perhaps the most striking feature apparent from this last case study is the strong community affiliation of the goshala. Here, in upper Assam over two thousand kilometers from its geographic and cultural heartland, the Marwari community founded, some three-quarters of a century ago, a goshala for the protection of cows which even today remains essentially a Marwari vania institution.

The preceding case studies provide a comprehensive view of the workings and functions of three goshalas in three different regions of India. If the relatively modern educational and Gandhian institutions suggest that the origins of the goshala are to be sought in the beliefs and values of ancient Indian society, this position is reinforced by the historical and functional dimensions of the temple institutions as illustrated by the Śri Nathji Goshala in Nathdwara. These values reappear in the cow-protection role of the vania institution as well, though here economic activities are of increasing significance. The goshala even appears to fulfill certain social functions in the community at large. It now remains to be seen whether the conclusions reached in the discussion of selected institutions can be applied with equal validity to the rest of India.

VII

Goshalas and Pinjrapoles in Modern India—Organization, Management, Community Support, and Finances

In an age when society is becoming increasingly complex, and when social, economic, and religious institutions are becoming ever more subject to governmental regulation, the animal homes of India have undergone changes during the last few decades in keeping with developments in the country as a whole. Despite this, in their day to day working, the goshalas and pinjrapoles of today combine certain aspects of medieval Indian society with elements of the twentieth-century economic and cultural scene. This is most apparent in the organization, management, and finances of the contemporary institutions.

Organization and Management

Just as there exist variations in the types of animal homes to be found in India, so there are differences in the way the institutions are organized and managed. There is, for example, the privately run goshala. Even though these institutions may be dependent on public donations, they are founded and personally managed by religious-minded individuals concerned with the protection of economically useless cattle. One such animal home, for example, is the Shyam Goshala near Vrindavan, founded in 1964 by Baba Bansiwala, a Rajput who has devoted his life to gosevā, service to the cow. Private institutions are usually small-scale operations, involving limited numbers of animals and somewhat tenuous funding. They rarely outlive their founders. Indeed, of the total number of animal shelters surveyed by this author, only three were goshalas run by private individuals. No doubt others exist, but as they have no formal associations with state goshala federations or government departments, records as to their numbers are not available.

A second category of institutions is affiliated with some educational or religious organization, such as the gūrūkul or temple. These are generally internally administered and thus, not being subject to external controls, also assume the character of private institutions. The temple, ashram, and educational goshalas, for example, are financed and managed through their respective parent bodies and fall into this grouping.

Most pinjrapoles and vania institutions, however, are today organized either as charitable trusts or as charitable societies registered under the various Societies Acts of the states of India. Most of the trusts have been founded by prominent citizens of religious or philanthropic bent and are located in the premier cities of India, but such individuals often establish goshalas at centers which have special religious significance for the Hindu. A religious man named Hasananda Verma, for example, desired that all of Vrindavan and Mathura, the land of Krishna's childhood, be devoted to the welfare of the cow. He traveled around the country collecting donations for this end, finally approaching M. M. Malviya, founder and first Vice Chancellor of Banaras Hindu University, and a group of wealthy Calcutta businessmen, among them the industrialist, Birla. These men were sympathetic to his cause and set up, in 1935, the Mathura-Vrindavan Hasananda Gocharabhumi Trust. Today the head office is in Calcutta, and the trustees, all of whom are Agarwal Vaishnavas, live in Calcutta, Kanpur, Mangalore, Bombay, and New Delhi, while the trust maintains its goshala at Vrindavan. Similarly, both the Śrimad Vallabh Goshala at Gokul near Mathura and the Udhaokund Goshala near Goverdhan are maintained by trusts from Bombay. The locations of these institutions assume special sanctity from their associations with the legend of Krishna.

By far the most common organizational form for the goshala and pinjrapole to take, however, is that of the registered society. Most of the institutions not falling under Public Trust Acts come under Registration of Societies Acts or, in states such as Uttar Pradesh and Rajasthan, legislation aimed specifically at the regulation of goshalas and pinjrapoles (Appendix B). Under the terms of most of these acts, institutions are required to appoint trustees or a board of officers, set up rules and regulations for the running of the society, and maintain regular accounts, which are audited annually and made available for public inspection.

Although the provisions of the various acts differ from state to state, certain benefits may accrue to an institution depending on its designation. In some states, for example, animal shelters, as charitable institutions, are exempt from income tax and from the terms of the Land Ceiling Acts being enforced as part of agrarian reform programs.

The system of management illustrated by the Rajkot Mahajan Pinjrapole and the vania goshalas discussed as case studies is typical of the rest of the country. The affairs of the institutions are handled by a managing committee, usually elected by the members of the society, which then appoints its officers—president, secretary, and treasurer. These appointments are generally for the period of one year and are purely voluntary. Officers receive no remuneration for their services. In the larger urban areas, the president—often a man of considerable standing in the local community—may fulfill an honorary role, merely lending his name and prestige to the goshala or pinjrapole, but in most instances he takes an active part in its affairs. A fulltime, paid manager is usually employed to handle the day to day running of the institution, especially when dairy herds are involved. This position may be a permanent one, but even though he may be experienced in handling cattle, rarely does the appointee have any formal training in animal husbandry.[1]

The membership requirements of goshala and pinjrapole societies vary throughout the country. In some cases, fixed membership fees are set. At the Mirzapur Goverdhan Goshala in Uttar Pradesh, for example, the annual fee (or donation) for an ordinary member is Rs. 11. One can become a life member for Rs. 101 and a patron (*sanrakshak*) by contributing Rs. 5,000. Usually, however, there are no set rules, and one becomes a member merely by making a donation to goshala funds. In this instance, the managing committee and officers are drawn from those members of the local community most actively concerned with the support and welfare of the institution, understandably so in view of the voluntary nature of the undertaking.

Community Support

An analysis of the composition of the management of goshalas and pinjrapoles reveals distinct community affiliations reflected in the support received from the public at large. It immediately becomes apparent that the animal refuges are organized, maintained, and patronized by specific social and religious elements within the community rather than by Indian society as a whole. Indeed, in response to my questions, many Indians professed total ignorance of the institutions. These communal associations are of considerable significance in that they provide valuable clues not only to the nature and functions of goshalas and pinjrapoles but also to their origins as cultural institutions.

The pinjrapole is clearly a Jain institution. This is apparent from the discussion of both distributions and the Ahmedabad and Rajkot pinjrapoles, and is further illustrated by other institutions in the country. Over the period of 1974–1975, data were gathered from twenty-one pinjrapoles in Gujarat, a sample of 13.9 percent of the institutions in the state and well over 10 percent of all the pinjrapoles in India. Of this number, fourteen had been founded by Jains or Jain organizations[2] and, though no reliable information is available for the remainder, it is highly probable that Jains were closely involved in the setting up of these, too. In eleven instances, the pinjrapoles received some sort of lag or other funding from Jain mahājans.

If the pinjrapoles are essentially Jain institutions dedicated to Jain ideals of ahiṃsā and jīv-dayā, they also receive considerable support from the Vaishnava vania community in Gujarat. During the nineteenth century, the predominantly Vaishnava Silk Dealers Mahajan in Ahmedabad ran their own institution, while Vaishnava support for the Ahmedabad Pinjrapole, and some of its related problems, has already been noted. Several factors can account for this, though they all seem to be related to a general acceptance of certain religious concepts, to basic similarities between the Jain and Vaishnava vania communities, and to the social and cultural history of Gujarat.

Although Jains have numerically always been a relatively insignificant element in the Indian population as compared with Hindus, the impact of their philosophies has far exceeded their actual numbers, especially in their stronghold of western India. Under the influence of Hemachandra (A.D. 1088–1173), for example, Kumārapāla, King of Gujarat, converted to Jainism, making it the state religion and banning the killing of animals throughout his kingdom. The Hindu inhabitants of Gujarat, therefore, have been long exposed to the beliefs of Jainism and would have little difficulty in accepting concepts such as ahiṃsā common to both religions. With the cult of Krishna and its associated respect for the sanctity of the cow also well entrenched in Gujarat, and with prolonged contact between the Jain and Vaishnava vania communities in their mercantile activities, a certain overlapping of common interests would be natural. As the pinjrapole provides refuge for all animals, including the cow, the Vaishnava can find in this institution a means of expressing his feelings for this sacred animal, and even of pursuing the ritual and practices surrounding it as prescribed by his religion. Conversely, though the goshala serves only cattle, the Jain views it as a specific application of the general principles of ahiṃsā and the sanctity of life, and thus worthy of support. This blurring of religious distinctions is especially noticeable in the smaller towns of Gujarat, where the mahājans become trade associations of mixed membership rather than caste organizations. The Viramgram Khoda Dhor Pinjrapole, some sixty kilometers west of Ahmedabad, for example, is a true pinjrapole—it maintains a kabūtriya and shelters pigeons, peacocks, hares, and horses as well as useless cattle and buffalo; yet it was founded and is managed by the city mahājan with both Jain and Hindu members. Similarly, the Visavadar Gorakshan Pinjrapole near Junagadh in Saurashtra was established by a Jain and a Pushti Margi for the purposes of both jīv-dayā and gorakshan, and is still supported jointly by the Jain and Vaishnava vanias of the town.

A third distinct community in western India that lends its active support to the pinjrapole is the Parsees. Zoroastrians who fled their homeland early in the eighth century to escape persecution by invading Moslems, these people settled along the western coast of India and subsequently came to be known as Parsees, after their land of origin, Persia (Fars). Although there are only some 100,000 Parsees in India today, they have remained a viable community, achieving prominence in the business world. Many leading Parsees have reputations as philanthropists, and pinjrapoles are among the institutions that receive their support. The Bombay Pinjrapole was started in 1834 by a Parsee, Sir Jamsetjee Jeejeebhoy, and it continues under the presidency of his great-great-grandson. This institution and others, such as the Poona and Surat Pinjrapoles, continue to receive the patronage of the Parsee community. The Parsee concern for animal welfare would appear to stem from the general regard for the sanctity of life found in the teachings of Zoroaster and in the sacred literature of the Zoroastrian religion.

The goshala, on the other hand, with its particular emphasis on the cow, is uniquely Hindu. Although specific attitudes and practices toward the cow may vary according to caste and sect, the sanctity of the cow pervades all of Hinduism. That the protection of the cow is an ideal common to upper caste Hindus other than Vaiśyas is seen at Vrindavan and Dibrugarh, where Brahmans serve on the managing committees of the goshalas. One of the local trustees of the Vrindavan-Mathura Hasananda Gocharabhumi Trust Goshala is Purushottam Goswami, a leading Brahman of the Chaitanya sect of Krishna worshippers and chief priest at the Radha Raman Temple in Vrindavan. The founder of the Shyam Goshala at Vrindavan is a Rajput of Kṣatriya descent. Support for the goshala, moreover, comes from peoples of many different persuasions within Hinduism other than Krishna worshippers. Adherents to *Sanātana*

Dharma[3] and to the Arya Samaj movement[4] find outlets for their beliefs in the sanctity of the cow and concern for cow protection in the goshala, being involved in both the establishing and maintaining of such institutions.

Nonetheless, as illustrated by the Vrindavan and Dibrugarh examples, the most overwhelming support for the goshala comes from the Marwari vania community. It is this group of traders and merchants, often speaking the Rajasthani dialect of their homeland, who have been the driving force in establishing and maintaining the cow homes of India. Largely Vaishnava or Jain by religion, and Vaiśya by caste, they would appear to be responsible for the presence of the goshala throughout the country. Both the Vrindavan and Dibrugarh goshalas typically reflect the extent of Marwari involvement with the institutions. Of the forty-five vania goshalas that I visited in the field, Marwaris were responsible for the establishing of thirty-one. Indeed, one Marwari, an Agarwal from Bihar, traced the origins of the institution back to the Agarwals themselves.[5] Marwari concern for the cow is also reinforced by the traditional strength of the Krishna cults in western India and among the vania community in general. Thus, even where the Marwari element may be lacking, local vanias continue to provide support and backing for the goshala.

Just as Vaishnavas support pinjrapoles in western India, so Jains support goshalas elsewhere. In fact, as the Jains become a numerically smaller element in the population away from their main concentrations in the west, they seem to lose their distinct identity and merge with the Hindu community. The Jains of Gujarat see themselves as separate from the Hindus, whereas Jains interviewed in small towns in Assam regard themselves as a part of the Hindu community and their contributions to the goshala as typical of Hindu reverence for the cow.

Although the goshala is characteristically a Hindu institution, it also finds acceptance in a non-Hindu community, but one that has embraced much of Hindu thinking. Sikhism, a syncretic religion combining the monotheism and egalitarian structure of Islam with basic Hindu religious concepts, has been presented as a revolt against the dominance of the Brahmans, caste, and the complexity of Hindu ritual. Yet one feature of Hinduism which did carry over into Sikhism was a respect for the sanctity of the cow, a respect which was no doubt reinforced by the traditional emnity between the Sikh and Moslem communities. According to L. N. Malhotra, Secretary of the Punjab Federation of Goshalas and Pinjrapoles, while it is uncommon for Sikhs to be actively involved in the running of goshalas, they do contribute to their upkeep, both in cash donations and in fodder, and will use the institutions to dispose of their useless animals. In the Punjab, however, as in most of India, the management of the institutions is largely in the hands of Marwari vanias.

Surprisingly, there are instances when Moslems, too, make use of goshala facilities. This should not, perhaps, be totally unexpected, since Moslem rulers in India historically have been known to promulgate cow protection,[6] but there is a considerable difference between official state policy over four centuries ago and the actual situation in India at the present time. Moslem farmers in the Punjab place cattle in goshalas, a practice also found in Gujarat. In Mandal, for example, a Moslem farmer of the Sunni sect reported that he had placed two oxen in the Mandal Pinjrapole because he was unable to maintain them under the existing scarcity conditions. An estimated sixty Moslem farmers from the region had animals in the pinjrapole and, though they did not contribute money to the institution, they donated grass for the feeding of the animals whenever possible. The management of the Dibrugarh Goshala in faraway Assam, however, noted that several unsolicited donations of cash had been received from Moslems during the preceding few years. There was even one institution included in my goshala survey, at Katni in the Jubbalpur District of Madhya Pradesh, that included a Moslem in its managing committee.[7]

Moslem involvement with the animal refuge in India is, understandably, minimal and seems to be related to factors other than religious. But there is one other religious community that is noticeable in its apparent lack of association with the animal homes of India. Like Hinduism and Jainism, Buddhism embraces concepts of ahiṃsā, reverence for life, and compassion for animals. It was reputedly the Buddhist Emperor, Ashoka, who established the first pinjrapole in his capital, Pataliputra, two centuries before the birth of Christ. Buddhists disapprove of animal slaughter in general and often are vegetarians,[8] yet not one single instance was discovered of a Buddhist being in any way associated with a goshala or a pinjrapole.

This would seem to be contrary to widespread Buddhist concern for animal welfare, at least as seen in Tibet. A Tibetan *lama,* resident in Darjeeling, claimed that it was not uncommon in Tibet for lamas and even monasteries to purchase land in the name of *tsethar,* the saving of life, and keeping on it a variety of animals such as goats, sheep, and yak. Similarly, Tibetans follow the practice of buying animals from butchers to prevent their slaughter.[9] The freeing of animals for religious purposes is common throughout the Buddhist world today.[10, 11]

This lack of Buddhist involvement with the animal homes of India is relatively easily explained. Unlike the Hindu, the Buddhist has no special veneration for the cow. Even though certain Buddhist-influenced

›oples appear to reject beef and raise more barriers against the aughter of cows than of other animals,[12] Buddha himself never specifi-ılly prohibited the slaughter of cattle or the eating of beef. The Buddhist eoples of India, moreover, tend to be found in the northern mountain one, beyond the plains cattle economy and the main concentrations of oshalas and pinjrapoles. And, though Buddhism is one of the rapidly cpanding religions of India today, in actual numbers Buddhists form an ısignificant element in the population of the country. Thus, in one of the reat ironies of history, in the land that conceived and nurtured one of ıe great religions of the world, that religion virtually is no more.

'inances, Property, and Assets

Traditionally, the goshalas and pinjrapoles of India have been char-able institutions supported by public contributions; and the various ypes of dharmada, which still constitute an important element in the nancial resources of the institutions (table 13), are modern survivals om past eras. The formal dharmada lag of the mahājans may even be aced back to the trade guilds of Mauryan times which collected dues om their members and distributed funds to various religious charities. ther lags, though perhaps of lesser antiquity, were used to raise funds r animal homes in the past[13] and similar features can be seen in India oday. At Sayla, for example, a special lag is levied at the *octroi* post, a ollection point for local taxes on goods in transit, on all goods passing hrough Sayla by road, and the proceeds are donated to the pinjrapole. In auhati, the goshala receives a tax known as *bilti* paid by residents of he city on goods arriving by train. In Agra, the goshala society prints pecial goshala stamps, depicting Krishna worshipping a cow, which are urchased by members of the Agra Byapur (Trade) Association and ffixed to sales receipts. Some institutions even employ workers to solicit lonations from travelers in local bus and railway stations.

If dharmada is an important source of income for goshalas and injrapoles, it also accounts for the greater proportion of the assets ccumulated by institutions throughout the country. In some instances, he founders of the animal homes themselves contributed the premises nd land for their newly created institutions. In others, they petitioned ocal *zamīndārs* (landowners) for gifts of land to support the animals n the goshalas and pinjrapoles. Over the years, moreover, these initial oldings have expanded through additional donations of land and roperty, and perhaps one of the more important sources has been the ulers of the formerly independent princely states of India.

Royal patronage is, of course, a feature of the past, but it accounts for

TABLE 13

FINANCES OF SELECTED GOSHALAS AND PINJRAPOLES[1]
1973–1974

Institution	State	Annual income (Rupees)	% Dharm.	% Rent	% Agric.	% Milk	% Cattle	% Dung	% Hides	% Govt.
Ahmedabad P.	Guj	1,624,132–25	19.48	17.66	21.85	0.17	–	0.93	–	36.49
Agra G.	U. P.	162,398–59	61.51	4.16	0.07	33.48	2.16	5.86	0.78	–
Aurangabad G.	Maha	3,500–00	0.10	28.57	–	82.28	8.57	12.85	–	–
Dehri-on-Son G.	Bihar	4,291–00	14.76	–	–	78.12	0.21	–	–	–
Dibrugarh G.	Assam	164,807–86	52.58	9.16	3.64	26.56	0.65	–	–	5.02
Gauhati G.	Assam	460,282–94	14.79	40.09	5.77	37.21	0.93	0.29	–	–
Ghazipur G.	U. P.	5,000–00	68.41	31.10	–	–	–	–	–	–
Jaipur P.	Raj	445,825–45	32.87	0.94	9.12	22.74	–	6.85	–	2.84
Jalpaiguri G.	W. Ben	130,000–00	25.38	5.38	10.78	46.15	0.31	5.00	–	–
Jetpur P.	Guj	95,200–00	5.25	25.21	21.01	47.27	–	–	1.24	–
Junagadh P.	Guj	208,652–85	61.13	5.38	4.61	17.16	–	–	0.16	6.01
Kanadvani P.	Guj	[illegible]	[illegible]	[illegible]	[illegible]					

Kathmandu G.	NEPAL*	68,000–00	80.88	–	–	16.18	–	1.90	–	–
Mandal P.	Guj	99,107–00	10.09	1.92	2.52	–	–	4.04	30.27	50.54
Nadiad P.	Guj	3,600–00	27.78	69.44	–	–	–	2.70	–	–
Naroda P.	Guj	15,000–00	31.47	65.60	–	–	–	–	–	–
Patan P.	Guj	260,251–00	32.12	19.73	11.61	6.86	0.43	4.18	–	–
Poona P.	Maha	250,000–00	4.00	–	2.01	72.12	16.41	2.61	–	0.80
Prantij P.	Guj	9,500–00	5.26	78.91	–	–	–	10.50	–	–
Saidpur G.	U. P.	10,000–00	30.14	–	–	49.17	–	3.58	–	–
Tenali G.	Andhra	15,000–00	66.67	6.51	–	6.66	–	5.74	0.21	–
Udaipur G.	Raj	8,013–98	41.54	9.76	–	44.27	1.43	–	–	–
Varanasi G.	U. P.	142,792–71	26.13	9.23	–	19.04	–	0.36	–	–
Viramgram P.	Guj	108,000–00	7.41	2.78	27.77	9.26	–	–	6.48	46.29
Wardha G.	Maha	46,000–00	3.12	37.50	–	45.12	6.25	5.26	–	–

[1] All of these institutions are included in the author's survey of goshalas and pinjrapoles in India. Data presented here were provided by the management or taken from annual reports.
*Country

much of the accumulation of land by goshalas and pinjrapoles. Thus, in a deed dated 11th Vad Chaitra in the Samvat year 1908 (A.D. 1852), the State of Bhavnagar granted outright the village of Chhapriali and its lands, previously leased to Anandji Kalyanji, to this Jain organization for the welfare of disabled animals, in commemoration of the death of the Mahārāja of Bhavnagar.[14] Today, this entire village and some 2,000 acres of grazing land continue to be devoted to the maintenance of the animals in the Chhapriali Pinjrapole. The hundred acres of land belonging to the Siwandi Gate Goshala in Jodhpur was donated by the Mahārāja of Jodhpur himself. In 1897, the Lal Ji temple goshala in Bharatpur received a grant of 14 acres of land from the Diwan's Office of the State of Bharatpur. Some 500 acres of grazing land at Merwera village belonging to the Pinjrapole at Palanpur were deeded to that institution by a former Moslem ruler of Palanpur State.[15]

Whether land and property were acquired from rājas, zamīndārs, and vanias, or whether they were purchased by the institutions themselves, these assets are an integral part of the financial resources of goshalas and pinjrapoles. Although much of the land is of poor quality and is used for grazing purposes only, many institutions farm cultivable land, growing both fodder crops for their own purposes and cash crops for sale on the open market. Production varies according to quality of soils, precipitation, and the availability of irrigation facilities, but agricultural income can account for as much as 20 percent of the total annual income of an institution. Even this figure does not provide a true measure of the value of land holdings, since fodder produced and used by an institution reduces the necessity for outside purchases, a feature which is not reflected in income statistics.

Similarly, the rental of land and buildings generates continuing income for goshalas and pinjrapoles, though again actual amounts vary according to the institution. Thus, smaller goshalas such as the Krishna Goshala in Ghazipur depend almost entirely on charitable donations, while others such as the Ahmedabad Pinjrapole Society or the Calcutta Pinjrapole Society receive as much as 20 percent of their not inconsiderable income from rents (table 13).

If dharmada, agriculture, and rent are major sources of financial support for goshalas and pinjrapoles, the cattle population of the institutions contribute to their own upkeep. Surplus milk is sold in the bazaar, and dung that is not used on goshala land or by goshala workers is sold as manure or fuel. Though institutions rarely sell female calves, males are sold to local agriculturalists, while working animals are sometimes made available for rent. Carcasses are usually buried or removed by Chamars, though occasionally contracts are offered for their disposal. In traditional institutions concerned with useless animals, minimal income is derived from these sources. However, where goshalas and pinjrapoles have become involved in commercial dairying and cattle breeding, the sale of milk and cattle assumes a much greater role in the finances of the institution.

An occasional source of financial aid available to goshalas and pinjrapoles is government grants and subsidies. As a part of India's national economic development planning, the second Five Year Plan included a Goshala Development Scheme, allocating funds to be administered at the state level for the improvement of goshalas and pinjrapoles throughout the country. Grants-in-aid and subsidies were made available to selected institutions for the purchase of quality-breed milk cattle, for the improvement of goshala facilities, and for the establishment of cattle development programs. In addition, during times of prolonged drought and famine, cattle relief funds such as those provided by the Gujarat State Government in 1973 may be made available to institutions involved in cattle protection.

The availability of government funds, however, is intermittent at best, and then is accessible only to a few institutions in the country. The financial mainstay of many animal homes in India continues to be the traditional dharmada, charitable donations made in the name of the religious concept of dharma, and income derived from accumulated land and property, supplemented in some instances by dairying and cattle breeding activities.

VIII

Animal Homes, Religion and Society—Ideals, Functions, and Festivals

Social institutions operate at several levels of reality and any attempt to analyze their nature and functions must take cognizance of this fact. There exists, for example, the ideal form of an institution—a model for accomplishing certain ends that are expressions of abstract or philosophical values. It is, moreover, these values that are often used to rationalize or justify the existence of the institution and that influence popular views and attitudes toward them. In contrast to this, there is the institution itself, set up to fulfill the stated objectives of the ideal form. It is characterized by a distinct organizational structure and modus operandi, and it engages in a variety of pursuits directed toward achieving its desired goals. These activities may also assume significance in ways not directly related to its basic objectives. In addition, various roles, some having nothing to do with intrinsic functions of the institution, are imposed on it from the outside.

These features are well illustrated by the animal homes of India. The detailed case studies of specific institutions suggest that they are conceptualized in religious or philosophical terms, but that they fulfill real economic functions in modern India and assume wider social attributes that have nothing to do with their animal-protection activities.

Goshalas and Pinjrapoles: The Ideal Forms

The protection and preservation of animal life in the animal shelters of India must be viewed not in terms of twentieth-century conservation, esource management, or economic development but must be placed in he context of those views of man and nature which have encompassed ndian life and thought for centuries past. The fundamental assumptions nderlying the religous philosophies of India are so different from those f the western tradition that they form a completely distinct frame of eference that must be understood before many "things Indian" can truly e appreciated. The existence of animal homes in India today, for xample, makes sense only in the context of the doctrines and dogma of he Indian religious tradition.

For the Hindu, the Buddhist, and the Jain, there exists no heaven or ell, no supreme all-powerful deity, no paradise full of beautiful *houris,* ut rather an unshakeable belief in the harmony of the universe. It is this asic unity of nature and the associated concepts of dharma, karma, and *aṃsāra* that provide the rationale for the acceptance of ahiṃsā and for ts expression in institutions such as the animal home. Specific interretations of these concepts and the relative values placed on ahiṃsā ary among the religions of India, however, and it is in Jaina philosophy hat the ahiṃsā concept finds its greatest acceptance.

The Jain religion is essentially atheistic for, while not denying the xistence of gods, it views the universe as limitless and eternal, functionng through the interaction of souls, of which there are an infinite umber.[1] The soul (jīva) is naturally bright, all-knowing, and blissful. As result of activity, however, karma adheres to the soul, karma being nvisaged as matter in fine atomic form. This subtle matter, invisible to he human eye, adheres to the naturally bright soul, which thus acquires irst a spiritual body and then a material one. In this manner, the soul ecomes imprisoned in the material world and embarks upon a cycle of ransmigration (saṃsāra) and reincarnation that continues indefinitely.

Every soul, however, possesses the potential to free itself of karma and o attain eternal bliss, the Jain equivalent of salvation. This process nvolves a lengthy time span and a virtually endless cycle of transmigraion, and indeed most souls have no hope of attaining the desired state of race. Ultimate release can only be achieved by dispelling accumulated arma and avoiding the attraction of new. The former can be accomlished by penance and the acquisition of merit (*pūnya*), the latter can be nsured by carefully disciplined conduct. As the amount of karma ossessed at the time of death determines the nature of one's next ncarnation, freedom from karma is essential for happy rebirths and the inal release of the soul from the cycle of transmigration.

Jain concern for ahiṃsā derives from the need to avoid the accumulaion of karma, from a general principle of nonactivity. The negative ature of the concept is apparent in the word itself. *Ahiṃsā* is derived from the Sanskrit verb meaning "to kill or damage;" the *hiṃs* form is th desiderative, while the *a*-prefix negates the meaning, so that *ahiṃsā* ha the literal sense of "renunciation of the desire to kill or injure."[2] Thi renunciation and many of the patterns of behavior that spring from i especially some of those that seem irrational to the western mind, ar thus logical extensions of Jaina concepts of nature and of the universe.

For the Jain, ahiṃsā is the greatest personal virtue. Its impact on Jai behavior, moreover, is intensified by the belief that souls are found i inanimate objects as well as in plants and animals. Stones, rocks, o running water are in possession of souls just as are men, but the soul are so bound by matter that they are not immediately detectable. Th kicking of a stone violates the principle of ahiṃsā and accumulate karma just as does murder. The stone can feel pain but the cries of th injured soul cannot be heard as it has no means of expression.

According to Jain philosophy, all objects are grouped into five catego ies depending on the number of senses they are thought to possess. Th highest class of souls is attributed with five senses and into thi grouping fall gods, men, and the higher animals, including cattle and th horse.[3] Souls of the second division lack the sense of hearing and includ certain insects. Creatures such as ants and fleas, which allegedly ca neither hear nor see, belong to the third category, while those possessin only the sense of taste and touch comprise the fourth. The last class c souls contains objects thought to have only the sense of touch, that is plant life, earth, fire, and water.

Jaina monks must practice ahiṃsā toward all five classes of souls resulting in a life of extreme austerity and rigor. The first of the five vow taken by the Jain ascetic is that of ahiṃsā. Five buttressing clause (*Pañca Bhāvanā*) help him to keep this vow. He must never walk o beaten paths in case he cannot see insects and treads on them (*Īry samiti*); he must be watchful in speech so that he never instigate quarrels (*Bhāṣā samiti*); he must see that the food he accepts as alm contains no insects (*Eṣaṇā samiti*); he must ensure that anything h receives or keeps for his religious duties has no insect life on i (*Ādānanikṣepaṇā samiti*); and at night when he is disposing of the re mains of his food or any other refuse, he has to be careful that no insec life is injured (*Pratisthāpanā samiti*). Monks are strict vegetarians, bein selective even about the plants they eat and sometimes carrying their vo of ahiṃsā as far as to refuse all food and to die of starvation. Masks ar worn over the mouth to prevent the inhalation of insects, water i strained before use, hair is pulled out by the roots to avoid harmin vermin—these are all aspects of Jaina ascetic life which spring directl from the ahiṃsā doctrine.

Life for the layman is somewhat less restrictive, yet it, too, is defined by the philosophical boundaries of Jaina views of the world. Of the nine categories of fundamental truth, the first is jīva—soul or life. To take life (*hiṃsā*) is the most heinous of all sins (*pāp*), and to abstain from killing (ahiṃsā) is the most binding moral duty. In the first of the twelve lay vows (*Prāṇātipāta viramaṇa vrata*), the Jain undertakes never intentionally to destroy a jīva that has more than one sense.

It is only in this context, therefore, that the pinjrapole becomes something more than a cultural oddity or curiosity. The institution assumes the role of a mechanism enabling the Jain to express basic tenets of his faith. The moral consequences of hiṃsā may be avoided by the placing in the pinjrapole of useless cattle, injured animals, birds, and insects—all classes of jīva whose destruction would violate the ahiṃsā principle. In this manner, the accumulation of bad karma may be avoided. Moreover, active support of the pinjrapole in the name of jīv-dayā is an effective means of acquiring great pūnya, or merit, through which accumulated karma may be dispelled. Thus the pinjrapole may be seen to be an institution fulfilling what is primarily a religious function.

The rationale behind the goshala, on the other hand, is different from that of the pinjrapole, because the institution is derived from the Hindu rather than the Jain religious tradition. Although the two are similar in many respects, one significant area in which they differ is in the relative emphasis placed on ahiṃsā. Although the doctrine of nonviolence is an integral concept in Hindu religious thinking, it does not occupy the central position that it does in Jain philosophy. The goshala, therefore, is an institution that must be viewed not so much in terms of a general respect for life or the consequences of violating ahiṃsā, but rather from the perspective of Hindu reverence for one animal, the cow.

The basic objective of the goshala is gorakshan—the protection and preservation of the cow. This raises the question as to why the cow should be singled out above all animals for unique treatment. When faced with this question, the Hindu is likely to put forward one or more of a variety of reasons—the sanctity of the cow, dharma, ahiṃsā, and the economic utility of the animal. The existence of cow homes in India, therefore, needs to be seen against the background of these many facets of Hindu attitudes toward the cow, rather than in the overriding influence of a single philosophical concept.

The emergence of the sacred cow concept in India has been examined earlier in this study; at this point, suffice it to say that by the Epic Period, around the fourth century A.D., the doctrine was well established in the religious literature of Hinduism. From that time onward, moreover, the association of the cow with ahiṃsā forms a persistent theme in Hindu canonical and non-canonical literature, so that today the orthodox Hindu can justify the gorakshan function of the goshala not only in terms of the specific sanctity of the cow but also in terms of the general doctrine of ahiṃsā and of the moral consequences of taking life.

The gorakshan role of the goshala in modern India receives further ethical and social sanctions from the Hindu concept of dharma. Like ahiṃsā, dharma is found in Jainism and Buddhism as well as in Hinduism, but its interpretation differs according to religion. In Jainism, for example, dharma is usually seen as the universal rule of nonviolence, the Eternal Law. In Hinduism, however, dharma is one of the Four Ends of Man (*purushārtha*), a concept fundamental to Hindu attitudes towards life and daily conduct, and indeed basic to all of Hindu culture and society. Man can seek salvation in four ways, through material wealth (artha), pleasure (*kāma*), spiritual liberation (*moksha*), and dharma, but the first among these is dharma, the pursuit of righteousness, duty, and virtue. Although most commonly applied to the organization of society into a hierarchical class structure (*varnas*), and of the life of the individual into stages (*aśrama*), the concept of dharma is comprehensive and all-pervasive and generally refers to "religiously ordained duties . . . morality, right conduct, or the rules of conduct (mores, customs, codes of law) of a group."[4]

Gorakshan, the protection of the cow, becomes a part of this pursuit of righteousness. Not only are the sanctity of the cow and ahiṃsā accepted tenets of Hinduism, but various rules and regulations defining correct behavior toward the cow and cattle are set out in Hindu literature. It is, moreover, the dharma of certain groups to protect cattle. In the *Bhagavad Gītā,* part of the epic *Mahābhārata,* whose message is of the supremacy of dharma, gorakshan is set out as one of the duties of the Vaiśya, and service as the lot of the lower castes:

> To till the fields, protect the kine, and engage in trade, (these) are the works of peasants-and-artisans, inhering in their nature; but works whose very soul is service inhere in the very nature of the serf.[5]

In itself, this would seem to reflect more the economic utility of cattle, the recognition of which is apparent even in the earliest of the Vedas, than notions of the sanctity of the cow; yet, placed in the broader context of dharma and in the acceptance of varna roles, it assumes an ethical character. Many a Hindu vania accounts for his support of goshalas both from the point of view of dharma as well as of the economic value of the cow.[6]

Thus, even though the goshala may assume differing forms in India oday, the raison d'être of the institution is gorakshan. If this desire to reserve and protect the cow originates in part from an appreciation of ts economic value, it also has historical roots in Hindu mythology, and t reflects ethical and moral concepts such as dharma and ahiṃsā which re fundamental to Hinduism itself.

oshalas and Pinjrapoles: Perceived Forms

In their ideal forms, therefore, the goshalas and pinjrapoles of India re essentially religious institutions. That the modern institutions are iewed in these terms by many who organize and support them is clearly een in attitudes and practices associated with goshalas and pinjrapoles n India today. Of the thirty-three pinjrapoles that I visited during field esearch in India, all claimed that they were founded for a religious urpose, specifically for the protection of animal life in the name of himṣā and jīv-dayā. The majority of these, moreover, were established uring the last hundred years, so they cannot profess to be institutions of reat antiquity with roots in cultural values long disappeared from the ndian cultural scene.

The current activities of some institutions, such as the Jain Bird Iospital in Delhi, continue to be in keeping with their original objectives, ut even where pinjrapoles have assumed a nonreligious character, istorical sources and documents substantiate claims for religiously nspired origins. Today, for example, the Bombay Pinjrapole has developed into something far more than a mere institution for preserving nimal life. It maintains a sizable commercial milk herd, is actively nvolved in cattle breeding and development, and markets its cattle over wide area of western India. Yet its founding purpose was to prevent, for eligious reasons, the indiscriminate slaughter of dogs. In a trust deed lated 18 October 1834 (see Appendix C), the merchants of Bombay greed to pay fees for the setting up of the pinjrapole so that the dogs on he island of Bombay, ordered killed by the East India Company, could be aved because "the killing of animals in general is inconsistent with the eligion of the Hindoos and the Parsees, and is a heinous act and equires atonement."[7]

Religious or ethical motivations are also consistently given as the eason for the existence of all but one of the sixty-five goshalas included n my field survey. Of this total, four were attached to educational nstitutions and three to ashrams, their purpose being to promote raditional Hindu attitudes towards the cow. Seven were temple goshalas, providing Vaishnava temples with milk for temple ritual and rasād, for the use of priests, and for feeding the poor. An additional institution, the Goverdhan Vilas Goshala in Udaipur, though not exactly a temple goshala, was supposedly founded by Rāna Udai Singh of Mewar to provide milk to Krishna temples in the city, a function that it still pursues today.

Of the remaining fifty institutions, most of which belong to the vania type, only one expressed the reasons for its founding in purely economic terms. The Gerua Gram Seva Goshala in Kamrup District in Assam was established in 1971 by S. K. Sharma, who calls himself a "private social worker," for the purpose of improving the economic conditions of local tribal peoples. Located in an isolated part of the state having a high proportion of Boro in the population, the goshala attempts to promote cattle keeping among these economically depressed peoples. As well as providing employment for local Boro, the goshala supplies milk, distributes seeds, and arranges demonstrations of agricultural and animal husbandry techniques. A scheme also exists whereby cows are given to local tribals. When the animal calves, if the recipient wishes to keep the cow and the calf, he will pay a previously agreed amount to the goshala. If he returns the animal, the goshala pays him the amount, to cover the cost of maintenance, but he is entitled to keep the calf. A Jersey cross and two deshi bulls are kept at the goshala for servicing the cows of local people free of charge. Although this operation is carried out on an extremely small scale (in 1974 there were only twenty-five animals in the goshala and only four heifers had been farmed out under this scheme) and is privately financed, it would seem to be aimed at introducing or establishing cattle keeping among Boro and other tribals in the areas. Today, the Boro keep cattle and use milk, but some thirty years ago the use of milk was uncommon. Though they do not worship the cow, the Boro claim to be of Kṣatriya caste and will not eat beef. It would seem, therefore, that the Boro are undergoing the process of Hinduization so frequently experienced by non-Hindu peoples on the margins of the Hindu culture area. Whether by accident or design, the Gerua Goshala appears to be furthering this process by attempting to establish a cattle-keeping economy among peoples who have traditionally been outside the economic, social, and religious rubric of Hinduism.

Some of the remaining goshalas covered in the survey included in their original functions cattle breeding and development (Arrah Goshala, Bhojpur District, Bihar), milk production (Śri Vyapari Gorakshan Sanstha, Virgaon, Nagpur District, Maharashtra), and protection of cattle from Moslems (Śri Krishna Goshala, Mathura, Uttar Pradesh). Yet all of them without exception expressed their primary role in religious terms—cow protection (gorakshan) and service to the cow (gosevā) stemming from dharma, the Hindu code of right conduct.

Thus, whatever their actual functions may be, the goshalas and pinjrapoles of India are perceived to be religious institutions. This point of view is further supported by the actions of many individuals who, though not necessarily associated with the organization and management of animal homes, make use of them in the course of their daily lives. Mention has already been made of the Jain vania family in Rajkot who visited the Rajkot Mahajan Pinjrapole to donate feed in the name of jīv-dayā in commemoration of the death of a relative, and indeed this is only one example of a whole pattern of ritualistic and religious behavior surrounding the animal home.

In Bombay, for example, anywhere between 200 and 400 people visit the pinjrapole every day, a volume that increases dramatically at the time of special festivals such as Makarśankrant, Dīwālī, Gopāṣṭamī, and Janamaṣṭamī, when visitors come to offer gopūjā. There are five days of the fortnight which are most favored for visiting the pinjrapole—*Ikam* (1st), *Aṣṭamī* (8th), *Ikadassi* (11th), *Purnima* (full moon), and *Amās* (no moon). A series of random interviews taken during the course of a morning at the Bombay Pinjrapole yielded the following biographical data and reasons for the informants' presence at the institution:

Informant 1. A Mehta, belonging to the vania (Vaiśya) community. By religion, the informant was a Pushti Margi, that is, a member of the Vallabhāchārya sect of Krishna worshippers. A steel merchant living in Bombay, he visits the pinjrapole from four to six times each month for purposes of dharma. On each visit, he purchases grain and fodder, these being offered for sale by the institution itself, and feeds the cows and the pigeons.

Informant 2. A Mehta, also a Vaishnava vania, who visits the pinjrapole on the 11th day (Ikadassi) of every fortnight in the name of dharma.

Informant 3. A Lohana, a Vaishnava by religion belonging to the vania community. A businessman living in Thana, twenty miles north of Bombay, he visits the pinjrapole every time he is in Bombay on business to offer dharmada and for darśan of the cows.

Informant 4. A Dhani, belonging to the Vaiśya caste and a member of the Pushti Margi sect. A government office worker living close to the pinjrapole, he visits the institution to offer dharmada every day before going to work.

Informant 5. An Alai Lohana, Vaiśya by caste and Pushti Margi by religion. An office worker who comes to the pinjrapole twice a month on Ikadassi for *godhan* (feeding the cows).

Informant 6. Lohanas (mother and daughter), Vaiśya by caste and Vaishnava by religion. They visit the institution every day to feed the cows because the husband tells them to!

Informant 7. An Agarwal vania (female), Vaishnava by religion. She comes to the pinjrapole to perform *Pakāma Yajña*, explained in the following manner. In the preparation of food, there will be killing of germs, and thus a violation of ahiṃsā. To atone for this, it is laid down in the *Laws of Manu* that every woman should perform Pakāma Yajña, putting aside five varieties of food to be given to cows. The informant visits the pinjrapole twice a day with food prepared in her kitchen to be fed to the cows!

These visitors to the Bombay Pinjrapole are representative of the many Hindus who attend goshalas and pinjrapoles throughout the country. Some, no doubt, go through the motions of mindless ritual without any real feeling or understanding, merely because it is customary or because they are told to do so. But for many others, the institutions play a meaningful role in daily rites and practices that are deeply rooted in the teachings and beliefs of their religion.

Religious Activities: Temples and Festivals

The religious associations of animal homes are reinforced by the presence of shrines and temples in goshalas and by the celebration of certain religious festivals in the institutions. Virtually all goshalas maintain shrines dedicated to Krishna, thereby underscoring the ties among the institution, the worship of the cow, and the cult of Krishna. Shrines may be relatively simple, set up by goshala workers and consisting of nothing more than a small statue of the god or, as at the Gokul Goshala, a stone taken from Goverdhan Mountain. If, however, they have been built by the goshala itself or donated by a wealthy patron, they may be much more elaborate or may even assume the form of a small temple. That the goshala has religious meaning for Hindus in general rather than specifically for the followers of Krishna is seen in the dedication of many goshala temples to Shiva and other deities. In the

Gauhati institution, one of the largest and richest in the country, there are three temples on the goshala premises, two dedicated to Shiva and one to Hanumān, the helper of Rāma. Shrines and temples are not, however, found in pinjrapoles, the Jain religion being essentially atheistic in nature.

Though many festivals of the Hindu calendar, such as Makarśankrant, Dīwālī, or *Holi,* may be observed by people visiting goshalas and pinjrapoles to make donations and for darśan of the cows, the main religious festivals celebrated in the institutions are Krishna-related. Gopāṣṭamī is celebrated in all goshalas throughout the country and is even observed in some pinjrapoles, a reflection of strong Vaishnava involvement with the institutions. Festivities may extend over a period of days and are usually similar to those described at Vrindavan and Dibrugarh. Goverdhan Pūjā, on the other hand, is more of a regional festival widely celebrated in areas of strong Vaishnava influence. Seeking the protection and multiplication of the herds of cattle, the rites of Goverdhan Pūjā as seen at Vrindavan center around the worship of the dung image of Krishna in his legendary defiance of the rain god, Indra.

Other festivals may be celebrated in goshalas, but these often reflect local or regional patterns of religious observances rather than any specifically goshala-oriented ceremonies. In strongly Vaishnava areas such as Rajasthan or western Uttar Pradesh, for example, Janamaṣṭamī (Krishna's birthday) and Holi are observed, but these rarely embrace more than the rites followed in the community at large.[8] In Maharashtra, several goshalas celebrate, in addition to Gopāṣṭamī, the *Polā* festival, an agricultural festival involving oxen rather than cows, occurring toward the end of the monsoon after the *kharīf* (autumn) crop is in the ground (fig. 20). The Gorakshan Sabha in Nagpur, for example, decorates oxen from the goshala with *ām* (mango) leaves and takes them from house to house, collecting funds for the institution.[9]

Another regional festival, celebrated mainly in Bengal, is Jagadhatri Pūjā (Śakti Pūjā), falling the day after Gopāṣṭamī and dedicated to the worship of the female principle, sometimes in the form of the "Cow Mother."

Occasionally other cattle-related ceremonies found in the community at large may be performed at goshalas. In Jodhpur, for example, people come to the goshala to worship cows at the time of *Vats Baras,* more commonly known as *Vashu Baras*, on the twelfth day of the dark fortnight of Asvin.[10] This day is dedicated to mothers with male offspring, both women with sons and cows with male calves. Women tell their children the myths and legends of Krishna, young children and calves are fed special foods, and pūjā is made to cows suckling male calves. Vashu

Fig. 20. *Bel Polā Festival in Maharashtra.*

Baras[11] and similar observances, however, are personal rituals that can be performed in goshalas rather than major events of the goshala religious calendar.

Religious Functions

If animal homes in India serve as the focus for a complex of ritual and religious activities, they also succeed in their basic religious function of preserving animal life. This is clearly seen in the case of the pinjrapole, for even though an economic or ecological rationale favors the protection of cattle, sheep, or goats, it certainly cannot account for the wide range of noneconomic animals given shelter in the pinjrapoles. The Ahmedabad and Rajkot institutions merely provide examples of the variety of wild and domestic creatures that are found in these asylums at the present time. Table 14 provides a summary of the animal-protection role of twenty-two pinjrapoles in Gujarat. The preservation of insects, birds, dogs, cats, peacocks, monkeys, and deer in these institutions assumes meaning only in the context of a social tradition which places great emphasis on ahiṃsā and respect for animal life in general. As recently as ten years ago, a pinjrapole in the Surendranagar District of Gujarat is reported to have had snake catchers who were called upon by householders to remove venomous reptiles from their homes, not to be killed but to be released in the jungle far from human habitation. Some of the animals, such as the monkey, pigeon, or peacock, may be accorded respect and reverence in keeping with their position in traditional folk culture and legend, but this alone cannot explain the regard in which all animal life is held. This is the outcome of a specific philosophy and, indeed, it is usual for those who support the pinjrapole to explain their actions in terms of ahiṃsā, jīv-dayā, and the gaining of pūnya or merit.

In the goshala, the distinction between the religious and the economic is less well defined, since gorakshan and gosevā both embrace the idea of cattle development as well as cow protection. The situation is further complicated by the existence of the various forms of the goshala, some fulfilling rather specialized roles. Yet all of the vania institutions from which data were gathered sheltered nonproductive cattle, totally useless animals that were too old to calve, or that were lame, blind, or otherwise incapacitated (fig. 21). Their continued survival can in no way be justified on purely economic grounds. They contribute nothing in terms of milk, calves, or traction, while the cost of maintaining them far exceeds the minimal returns derived from the sale of their dung and carcasses. The presence of these animals in goshalas, therefore, can only be viewed as a measure of Hindu reluctance to kill the cow, a reluctance

TABLE 14

ANIMAL PROTECTION ROLE OF PINJRAPOLES IN GUJARAT

	Commercial dairy herd	Jīvat Khān	Pārabaḍī/Kabūtriya	Kutta-kī-Roṭī Fund	Useless																All animals accepted formerly
					Cattle	Buffalo	Sheep	Goats	Horses	Asses	Poultry	Dogs	Cats	Birds	Peacocks	Hare	Deer	Nilgai	Camels	Monkeys	
Ahmedabad	x	x	x	x	x	x	x	x			x	x	x	x							
Bhavnagar	x		x		x	x	x	x				x		x							
Baroda					x	x	x	x	x	x											
Broach	x				x	x															x
Chhapriali					x	x	x	x	x							x	x			x	
Chotila			x	x	x	x	x	x						x							
Dakor	x				x	x			x												x
Idar	x		x		x	x			x					x						x	
Jetpur	x				x	x	x	x	x												x
Junagadh					x	x	x	x	x					x							
Kapadvanj					x			x	x												
Mandal			x	x	x	x	x														x
Nadiad		x	x	x	x	x		x						x							
Naroda			x	x	x	x															
Palanpur		x	x	x	x	x		x						x							
Patan			x		x	x		x	x												
Prantij			x		x	x		x				x									x
Rajkot			x	x	x	x	x	x			x	x	x	x	x						
Sayla					x	x	x	x				x		x				x	x		
Surat	x		x		x	x	x	x	x												x
Viramgram	x		x		x	x		x						x	x	x					
Visavadar	x				x																x

that stems more from considerations of the sanctity of the cow, dharma, and ahiṃsā than from an appreciation of the economic value of cattle.

The preservation of useless cattle, however, represents only one of the religious functions of the goshala in India. A more specialized but nonetheless important function is fulfilled by the temple goshala. The herds attached to temples are not sacred in themselves but are dedicated to the service of the gods, providing milk for temple ceremonies. Indeed,

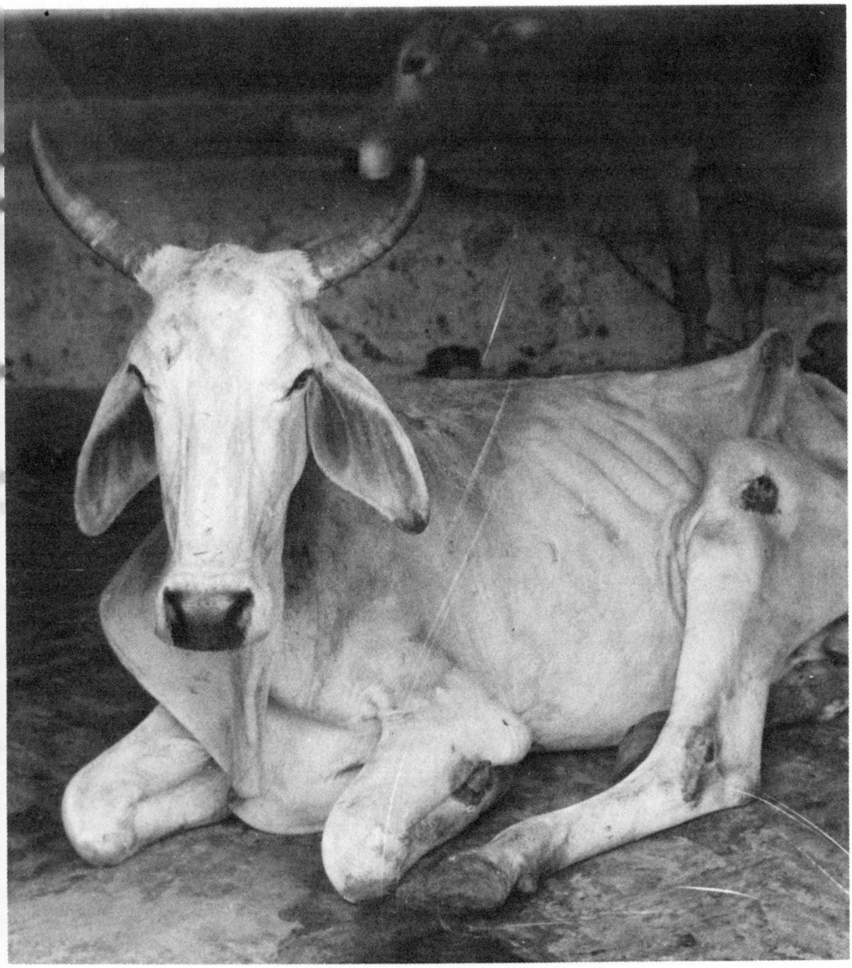

Fig. 21. *Useless cow in Sodepur Goshala*

milk and various milk products are perhaps the most important items used in Vaishnava temple ritual, both in daily routine and on special occasions. Details concerning the use of milk in temples are found in the *Hari-Bhakti-Vilāsa* of Gopāla Bhaṭṭa Goswāmī, a definitive compendium of Vaishnava rituals and practices followed at Krishna temples throughout the country. Thus, at the time of Mangāla, the morning levee, "The idol should be bathed with milk, curd, ghee, honey and sugar separately in that order and poured through a conch."[12] The *Hari-Bhakti-Vilāsa* defines the quantities that should be used for the *abhiseka* (bath) and specifically prohibits the use of buffalo milk for the ritual.[13] The glory of bathing the idol in pañcamṛit is sung in many of the *Purāṇas.* The *Agni Purāṇa* says that it is equal to donating one hundred cows.[14] The *Vishnu Dharmottra* tells us that it removes all mental and physical sufferings.[15] According to the *Narsiṃha Purāṇa*, by bathing the deity in milk one dispels all sins and by bathing it in curds one attains Vaikuṇṭha (the heaven of Vishnu).[16] In the *Skanda Purāṇa*, it is said that by bathing Vishnu with milk one gets the results of the *Aśvamedha* (horse) sacrifice, and that this increases tenfold with curd, and with ghī ten times the result of the curd bath. "If one performs the bath of Vishnu without any motive, he attains *mukti* (liberation)."[17]

Though the abhiseka is a part of the daily ritual in Krishna temples, it also plays an important role in other ceremonies. The consecration of a new idol involves bathing it in both pañcagavya and pañcamṛit.[18] On the *jayantis* (birthdays) of the different incarnations of Vishnu, *salagrāma* (a black crystal found only in the Gandaki River in Nepal and taken to represent the various incarnations of Vishnu) is consecrated in pañcamṛit. On Janamaṣṭamī, the birthday of Krishna, the temple idols are bathed in pañcagavya and pañcamṛit, to the accompaniment of the appropriate Vedic mantras.

In addition to the bathing ritual, the offering of food (*bhoga*) to the deity forms an important part of the daily routine at Krishna temples, and this food takes the form of milk and milk preparations—milk, curd, ghī, malai, luddus, and a neverending list of sweetmeats.[19] At night, just before retiring, the deity is served with sweet boiled milk only, a time known as *Dudha-Bhoga* (the feeding of milk), which has been the inspiration for hundreds of songs by the poets of Vraj.

Although the main role of the temple cows is religious in nature, that is, the providing of milk for temple ritual, they also serve the general needs of temple organizations. As at Nathdwara, the temple kitchens utilize dung from the goshala for fuel and use surplus milk for feeding the many pilgrims who visit this important shrine from all over Gujarat and western India. This is true of the other temple goshalas I visited, especially those attached to major pilgrimage centers. Tulsi Shyam, for example, is a Krishna temple located deep in the Gir Forest of Saurashtra, far from any major settlement. Pilgrims to the temple often have to stay overnight, and lodging and food is provided by the temple organization. Thus all the surplus milk from the goshala not required for temple ceremonial is used by the priests and visitors to the shrine.

Similarly, at the Jagannath Mandir in Ahmedabad, free milk amounting to some 2½ maunds (94 kg.) per day is distributed to the poor; in addition, meals are served twice daily at the temple to three or four hundred needy people, as well as to the several hundred *saddhus* (holy men) who bivouac in the vicinity of the temple itself. The temple goshala cannot meet all the requirements for this charitable distribution of food, but once again surplus milk from the goshala is used for this purpose.[20]

If temple goshalas fulfill specialized functions in providing milk for temple ritual, for the feeding of pilgrims and for the charitable distribution of food to the needy, they also share the ideal of cow protection. The temple goshalas, like the vania institutions, maintain useless cattle, but rarely are these accepted from the public; they are mostly the goshalas' own milk cattle, supported beyond their productive years until their death. The nature of the institutions is such, however, that they are primarily concerned with milk production, not for economic uses but rather for the performance of ritual and ceremony surrounding the worship and adulation of the gods of Hinduism.

Social Functions

As communal institutions, the animal homes of India assume roles having nothing directly to do with their religious character or their economic activities, serving rather as means for social action. Note has already been made of the incident in Ahmedabad during the early years of the nineteenth century concerning the allocation of charitable funds, in which the Ahmedabad Pinjrapole became the focus of a communal dispute between Jain and Vaishnava vanias. Other such occurrences are on record. The Bombay Gazetteer, for example, provides the following narrative account of a sectarian dispute in Viramgram, a town near Ahmedabad:

> We confectioners are nearly all Vaishnavs but nevertheless we used to shut our shops on Shravak holidays. One year the *pachusan* of the two sections of the Shravaks, the *Dasa* and the *Visa,* fell on different days and we said we would only observe that of the Dasa Shravaks. The next year we grew bolder and

> declined to observe any of the Shravak holidays at all as they declined to keep ours. The quarrel went on for some time and, aggravated by an attempt the Shravaks made to bring in some outsiders, our guild passed a law that no member should have any dealings with a Shravak. Thereupon two members who happened to be Shravaks seceded. We then besought the Vaishnav members of the merchants' guild to help us, and they, seeing our case to be just, stopped their contribution to the animal home and threatened to form themselves into a separate guild. This brought the Shravaks to terms and they agreed to keep one of our holidays.[21]

Thus the pinjrapole came to be used as a weapon in the communal discord that periodically ruffled the surface of mercantile life in western India.

The institutions continue to figure occasionally in local politics. At a goshala in a major city in Bihar, for instance, attempts during 1974 to oust members of the managing committee by an opposing faction were motivated by a dispute among the vania community at large. Similarly, the president of a goshala in West Bengal reports that failure to support the institution by members of the vania community leads to retaliation through the imposition of social sanctions. Weddings are boycotted and business is withheld until the recalcitrant member pays his allotted dues.

Generally, however, goshalas and pinjrapoles serve more positive social functions. They act as a unifying focus of interest in the communal life of Marwari and Gujarati vanias across India, communities that have maintained a distinct cultural identity, one aspect of which embraces a special regard for the sanctity of life. Support for the institution, moreover, becomes a matter of social status and prestige. It is not enough to have contributed to a goshala or a pinjrapole; it must be seen that one has contributed. Thus donors' names and contributions are published in annual reports, the more important sometimes being inscribed on walls and buildings. At the Ahmedabad Pinjrapole, special plaques are erected in the name of donors who have contributed more than Rs. 1,100. Many pinjrapoles maintain special boards recording the names of those who have offered tithi donations. This public acknowledgment of charitable contributions made in the name of dharma not only enhances the donors' social and economic standing but also makes visible the strength of their commitment to traditional religious values.[22]

Goshalas and pinjrapoles provide an outlet, a means of expression, for those who adhere to ancient beliefs and customs, but their influence is not restricted to the vania castes. Through fairs and festivals they bring the Great Tradition into the lives of the common people. Many of those who attend goshala celebrations, who follow Krishna and Balarāma through the streets with their herds of cattle, and who watch the Krishna

Līlā are illiterate peasants participating in their own great cultural heritage. In this respect, the animal homes of India are more than just religious or economic institutions; they are part of the complex system by which elements of Indian culture have been preserved and transmitted from generation to generation and by which traditional Hindu beliefs and values have survived the ravages of passing time.

IX

Economic Activities, Functions, and Ecological Mechanisms

Although animal homes have fulfilled a variety of religious and social roles in India, increasing emphasis is being placed on economic functions. In the case of the goshala, this merely represents an extension of the traditional concept of gorakshan, but the development of economic activities in the pinjrapole introduces an element not encompassed in the original ideals of the institution. Yet in recent years, in addition to and sometimes at the expense of their broader life-giving roles, several pinjrapoles have evolved into goshalas in all but name, even to the extent of developing commercial dairy herds.

A prime example of this process of change is seen in the Poona Pinjrapole. Founded in 1854 by the merchants of the city, in its organization and activities the institution was typical of the nineteenth-century pinjrapole.[1] After 1903, however, the pinjrapole entered a period of decline. Its main source of revenue, the cesses, were discontinued and charitable donations were decreasing, so that the activities of the institution were increasingly restricted until 1939, when they were completely reorganized under the auspices of a new secretary, Āchārya Shankar Ganesh Nawathe. A "progressive and rational cow protectionist," Nawathe felt that goshalas and pinjrapoles should not only protect unserviceable and disabled animals but should undertake cattle development along modern and scientific lines. He began implementing his ideas in Poona by segregating useless cows and upgrading serviceable animals by crossing them with the Gir bull, the Gir being one of the superior milking breeds of India.

This marked the first step in the transition from pinjrapole to goshala, a process that was completed in 1940 when animals other than cattle were no longer afforded asylum in the institution. At this time, moreover, the Poona Pinjrapole laid the foundations of its commercial dairy herd. During World War II and its immediate aftermath, shortages of milk and food were universally felt in the large towns and cities of India. To meet this problem in Poona, an association known as *Anna Samiti* was formed to distribute food and milk to the needy. The Samiti purchased a herd of Gir cows, entrusted them to the keeping of the pinjrapole, and distributed their milk free at various centers in the city. At the termination of this milk scheme, the pinjrapole acquired these cows and used them as the basis for the current milk herd, which numbers over one hundred animals.

Today, the Poona Pinjrapole serves only cattle. There is a gosadan section for nonproductive animals, numbering twenty cows in August 1974, but the main activities of the institution center on its Grading and Gosamvardhana sections. In the former, useful cows of nondescript breed are upgraded by crossing them with Gir bulls, while the latter, operating under the Maharashtra State Goshala Development Scheme, contains the pinjrapole's milk herd of Gir cows. Annual milk production amounts to 120,000 kg. and receipts from the sale of milk and cattle account for some 88 percent of the institution's annual income. Thus the Poona Pinjrapole, originally set up to protect stray dogs and cattle, has partly through circumstance and partly through design become a goshala with a strong emphasis on commercial activities.

This process of evolution has occurred consistently in pinjrapoles in the larger cities outside of Gujarat. In Bombay, Calcutta, Ahmednagar, and Nasik, pinjrapoles founded to protect animal life in the name of ahiṃsā have become institutions in which economic activities predominate over religious ones. Some have even developed into important centers of cattle breeding. The Nasik Panchayati Pinjrapole, for instance, in addition to sheltering useless cattle, serves as a subcenter for the Bharatiya Agro-industries Foundation at Uruli Kanchan, working on improving milk yields using crossbreeding and artificial insemination methods.

Economic Activities

The economic activities of goshalas and pinjrapoles relating directly to cattle keeping may be considered under the headings of dung, hides and carcasses, agriculture, milk production, and cattle breeding.

Dung. In keeping with the general pattern of its utilization in India, as much dung as can be reclaimed in goshalas and pinjrapoles is restored to the energy system in the form of manure and fuel. One institution, the Tulsi Shyam Mandir, even uses dung from its goshala to provide lighting for the temple complex, converting it into methane in a gobar gas plant. Dung not used by the animal homes themselves is sold on the open market, and thus a certain amount, however minimal, enters the local agricultural and household economy. Admittedly, actual financial returns from the sale of dung may be negligible,[2] but this involves no additional input or investment on the part of the institution beyond the effort of collection. It also makes available to the local community quantities of scarce fuel and organic fertilizer.

Not all the dung produced by goshala animals is reclaimed. In institutions that keep milk herds of any size, it is customary to segregate the useless animals from the rest. Whereas milk cows are kept on goshala premises or grazed on goshala land, the nonproductive animals are often released to graze along the roadside or even in surrounding jungle areas, where much of their dung is lost to the institution. Such wastage is greater with smaller goshalas, which have little land of their own and which turn their entire herds out to graze where they can. Even so some of this dung is recovered by harijans and low-caste peoples who collect it for their own uses. No visitor to India can fail to notice the numerous women of all ages who scour the streets and roadsides gathering the droppings of passing cattle.

Hides and Carcasses. Although full use is made of dung in goshalas, the same cannot be said for the utilization of hides and carcasses. These represent another by-product of the institutions' activities in sheltering cattle, but one whose economic potential is lost to goshalas as a result of their means of disposal. The quality of hides tends to be inferior, since cattle in the institutions die from disease, old age, or the general effects of malnutrition, and few goshalas attempt to capitalize on the number of carcasses that pass through their hands (table 13). Only 5 percent of all reporting institutions sell carcasses to tanneries for processing. In the remainder, dead animals are buried, left out in the fields where they are devoured by vultures and pariah dogs, or are removed by Chamars, who utilize the carcasses and hides for their own purposes. Sometimes the dead animals are donated to them, sometimes the Chamars pay a nominal fee to the goshala, while occasionally the goshala itself will pay for the removal of the dead animals. In a sense this practice maintains the traditional character of Hindu society, reinforcing the functioning of caste and also continuing to provide Chamars with a means of livelihood. Yet the situation in India is changing. The Poona Pinjrapole, for instance, reports that whereas in years past the local Chamars would dispose of carcasses, using the meat and hides, they have now converted to Buddhism in an effort to improve their social status, forswearing the eating of meat and even the touching of carcasses.

Agriculture. Over the years, most goshalas and pinjrapoles have acquired considerable acreages of land[3] with the purpose of providing grazing and fodder for their animals (table 15). Agriculture, therefore, forms a major aspect of the institutions' activities. Not all of the land is cultivable, with only some 7.45 percent being devoted to fodder crops, though agricultural land is often rented to tenants on a cash or share-cropping basis.

Farming practices conform to those prevalent throughout the subcontinent, with the monsoon-fed kharīf (autumn) crop providing the main harvest of the year. Many institutions (51 percent of the data sample), however, have developed irrigation facilities, which not only permits the growing of a *rabī* (spring) crop but also reduces dependence on the vagaries of the monsoon. The main crops grown for fodder in the drier north and west are millets, being replaced by rice in the wetter areas of the lower Ganges valley and Assam. Jowar (*Sorghum vulgare*) and bajra (*Pennisetum typhoideum*) account for most of the acreage under cultivation, while maize, wheat, pulses (*dan*), and sugar cane (*ganna*) are of lesser importance. Sometimes crops such as jowar are grown solely for fodder purposes, but usually the grain is separated and only the stalks are fed to the cattle. This practice does, of course, reduce the nutritional value of the fodder. In the predominantly rice-growing areas of Bihar, West Bengal, and Assam, virtually all goshala land under cultivation is devoted to rice, with paddy straw being the main source of cattle feed. Here, with higher precipitation and a longer rainy season, grasses such as napier grass, guinea grass, and lucerne are also grown to provide green fodder. A few institutions scattered around the country devote a small percentage of their land to the cultivation of cash crops such as jute, cotton, or vegetables, but the total acreage involved is negligible.

Despite their agricultural activities, in very few instances are goshalas and pinjrapoles totally self-sufficient in fodder,[4] and excess requirements have to be made up through purchases on the open market. Some additional fodder is received in the form of donations, but the cost of fodder and other cattle feeds such as concentrates remains the largest single operating expense incurred by goshalas and pinjrapoles. An institution such as the Darjeeling-Siliguri Goshala in Siliguri (West Bengal),

TABLE 15

LAND ASSETS AND FODDER REQUIREMENTS IN
SELECTED GOSHALAS AND PINJRAPOLES
1973–1974

Institution	State	Total land (acres)	Grazing	Cultivated		Est. fodder consumption (maunds)	Purch. of fodder (rupees)	Annual % expenditure on fodder
				Total	Irrig.			
1. Ahmedabad P.	Guj.	4,860	3,225	1,140	–	150,000	626,729–20	38.48
2. Agra G.	U.P.	25	10	10	10	11,000	101,208–24	62.32
3. Aurangabad G.	Maha.	125	125	–	–	75	1,000–00	28.25
4. Dehri-on-Son G.	Bihar	0.3	–	–	–	100	1,500–00	34.95
5. Dibrugarh G.	Assam	130	18.75	–	–	12,000	125,000–00	38.02
6. Gauhati G.	Assam	400	368	20	10	18,000	60,000–00	15.38
7. Ghazipur G.	U.P.	6.25	3.75	2	2	75	700–00	11.66
8. Jaipur P.	Raj.	127.5	96	30	–	15,000	255,000–00	55.94
9. Jalpaiguri G.	W.Ben.	120	15	90	–	13,000	20,000–00	15.38

10. Jetpur P.	Guj.	300	220	80	12	18,000	35,000–00	41.18
11. Junagadh P.	Guj.	400	400	–	–	18,000	11,879–22	5.69
12. Kapadvanj P.	Guj.	38	2	35	–	6,000	20,000–00	64.08
13. Kathmandu G.	NEPAL*	0.62	0.25	–	–	–	37,000–00	73.71
14. Mandal P.	Guj.	500	125	375	4	120,000	400,000–00	87.21
15. Nadiad P.	Guj.	6.25	–	–	–	525	2,100–00	80.76
16. Naroda P.	Guj.	0.1	–	–	–	–	–	–
17. Patan P.	Guj.	6,475	4,324	–	135	5,600	13,414–00	3.94
18. Poona P.	Maha.	150	130	20	20	10,000	120,000–00	48.00
19. Prantij P.	Guj.	–	–	–	–	600	10,000–00	78.23
20. Saidpur G.	U.P.	5	1	4	4	200	20,000–00	76.92
21. Tenali G.	Andhra	1	–	0.5	0.5	900	15,000–00	90.00
22. Udaipur G.	Raj.	3.75	3.75	–	–	150	3,305–54	29.69
23. Varanasi G.	U.P.	120	10	100	50	15,000	46,745–31	32.72
24. Viramgram P.	Guj.	1,119	819	300	–	19,000	90,000–00	83.33
25. Wardha G.	Maha.	250	50	200	–	3,650	10,950–00	23.80

*Country

which has 28.12 acres of grazing but no cultivable land,[5] is forced to buy all its fodder to feed some 360 animals. In 1973, this amounted to 606.26 short tons of paddy, maize, bajra, and concentrates purchased at a cost of some Rs. 60,000. In many cases, feed costs may run to over 75 percent of total outlays. The Tenali Goshala in Andhra Pradesh, for example, owns no grazing land, and 90 percent of its annual expenditure is devoted to the purchase of feed (table 15). Thus, even though animal homes are producers of fodder crops, they are also major consumers, competing on the market for available supplies.

One offshoot of goshalas' agricultural activities worthy of brief mention is their input into the local agricultural economy beyond the renting of land or the sale of grain. Some institutions make available for hire their farm equipment, such as tractors, ploughs, and even working bullocks. The overall impact of this is minimal, but since much of the work in the fields has to be undertaken within a limited span of time, usually at the onset of the rains, and since shortage of bullocks for traction is a perennial problem throughout much of India, this practice can be of considerable value to the local farmer.

Milk Production. Although dung and hides are merely by-products of the normal activities of animal homes and offer limited potential for economic exploitation, the production of milk offers considerable scope for expansion. Not all cattle placed in goshalas and pinjrapoles are totally useless, and it is common practice to salvage animals for their milking capabilities, even though this may be low in the case of deshi cows. Nonetheless, investment on the part of the goshala is minimal, since institutions not keeping stud bulls have access to government animals for breeding. From this it is but a small step to developing and upgrading a milk herd on a commercial basis, especially in light of government-sponsored programs to provide quality milk cows and stud bulls as well as financial assistance and technical expertise to institutions in order to develop the cattle resources of the country.[6] Indeed, the management of several goshalas and pinjrapoles has expressed the view that, at a time of declining public support, this is the only way institutions can remain economically viable, supporting their useless animals on income generated by the useful.

Once a milk herd has been established in a goshala or pinjrapole, it perpetuates itself since female calves are rarely sold or given away. Of all the institutions surveyed that are not specifically involved in cattle breeding, only two report that they make available female calves to the public. All the others keep the female, adding them to their own milk herds, although they dispose of male calves, either selling them or giving hem to local farmers for agricultural purposes.[7] In this sense, goshalas nd pinjrapoles contribute to maintaining the numbers of bullocks in the gricultural sector of the economy.

Though milk herds are a common feature of animal homes throughut India, not all institutions have placed much emphasis on their levelopment. Four goshalas and five pinjrapoles in the study sample had no cows in milk or even between lactations, while in another twentywo the number of useless cows exceeded milk cows, suggesting that the ow-protection aspects of the institutions had not been entirely subordinated to the economic (table 16).[8] Even where milk cows were mainained, their productivity was low. Daily yields from deshi cows rarely xceeded two kilograms per day and in some institutions were less than one.[9]

Many of the larger and better endowed institutions, however, have acquired high-quality milk breeds around which to build their milk herds. The Vrindavan Panchayati Goshala, for example, bought twenty Haryana cows in 1974, and its milk herd numbered 102 animals at the end of that year. In addition, two Jersey bulls were purchased from the National Dairy Research Institute's cattle farm at Karnal for crossbreeding purposes. Daily milk yields averaged 2.69 kg. per animal in 1974, with a total annual production close to 99,000 kg. Larger institutions, such as the Calcutta Pinjrapole Society with annual milk yields ranging over 300,000 kg., make valuable contributions to the urban milk supply.[10] Milk production in many goshalas and pinjrapoles is being improved through the upgrading of deshi cattle, and by the introduction of better Indian milking breeds such as Gir, Haryana, and Sahiwal, and even of western dairy cattle, with many goshala herds being managed along the lines of modern dairies.[11]

Cattle Breeding. Though many goshalas and pinjrapoles regularly make available to the local agricultural population male calves and even, on occasion, female calves or milk cows, a handful of institutions with the necessary leadership, resources, and technical ability have expanded their activities to include cattle breeding on a commercial basis.

Today, for example, though still maintaining a gosadan section with some 1,000 head of useless cattle, the Bombay Pinjrapole also runs a milk herd of over 400 Gir cows. In addition to its milk production, however, the institution sells cattle to the government and to the general public as well. Table 17 shows sales of cattle to various organizations over the period 1970–1975. Most of these were effected in Maharashtra and neighboring states, with government departments and institutions being the main purchasers. Private individuals also purchase animals

TABLE 16

BOVINE POPULATIONS AND MILK PRODUCTION IN SELECTED GOSHALAS AND PINJRAPOLES 1973–1974

Institution	State	Total bovines	Total cattle	Cattle: Cows in milk	Cattle: Dry milkers	Cattle: Useless cows
1. Ahmedabad P.	Guj.	2,126	2,114	3	2	681
2. Agra G.	U.P.	384	384	41	106	130
3. Aurangabad G.	Maha.	35	35	6	–	29
4. Dehri-on Son G.	Bihar	23	23	3	7	3
5. Dibrugarh G.	Assam	270	270	40	60*	50
6. Gauhati G.	Assam	495	495	114	37*	6
7. Ghazipur G.	U.P.	19	19	–	2	12
8. Jaipur P.	Raj.	430	430	36	46	277
9. Jalpaiguri G.	W.Ben.	223	223	45	38	8
10. Jetpur P.	Guj.	304	298	27	53	25
11. Junagadh P.	Guj.	187	186	27	35	28
12. Kapadvanj P.	Guj.	81	81	–	–	81
13. Kathmandu G.	NEPAL**	69	69	7	20*	12
14. Mandal P.	Guj.	1,708	1,509	5	201	587
15. Nadiad P.	Guj.	27	25	–	–	25
16. Naroda P.	Guj.	–	–	–	–	–
17. Patan P.	Guj.	687	552	15	446	23
18. Poona P.	Maha.	189	189	50	59*	19
19. Prantij P.	Guj.	33	32	–	–	4
20. Saidpur G.	U.P.	78	75	3	7	37
21. Tenali G.	Andhra	29	29	2	2	22
22. Udaipur G.	Raj.	26	26	1	8	–
23. Varanasi G.	U.P.	132	131	38	20	33
24. Viramgram P.	Guj.	398	299	6	7	151
25. Wardha G.	Maha.	178	178	11	37	55

*Includes western crossbreeds
**Country

from the pinjrapole, and again they come mainly from neighboring areas, though sales have been made as far away as Bihar (map 11).

For individual purchasers, an application form requesting details of landholdings and so on, a reference from the gram panchayat (village council), and a local Bombay reference has to be submitted to the pinjrapole. The pinjrapole inspector contacts the local referee and verifies the

TABLE 16—Continued

	Cattle							Annual milk yields (kg.)
	Breeding bulls	Other bulls	Working bullocks	Useless bullocks	Female calves	Male calves	Buffalo	
1.	2	–	16	683	321	406	12	1,345
2.	5	–	3	–	67	32	–	36,500
3.	–	–	–	–	–	–	–	1,440
4.	3	–	–	–	6	1	–	2,920
5.	9*	–	6	–	63	42	–	40,000
6.	18*	–	4	–	201	115	–	122,338
7.	1	–	3	1	–	–	–	–
8.	4*	–	16	–	30	21	–	49,604
9.	8	–	11	2	49	62	–	82,280
10.	3	2	15	11	122	40	6	40,000
11.	–	–	–	41	37	18	1	20,460
12.	–	–	–	–	–	–	–	–
13.	4*	–	–	–	14	12	–	1,825
14.	4	–	2	497	117	97	199	–
15.	–	–	–	–	–	–	2	–
16.	–	–	–	–	–	–	–	–
17.	–	4	28	18	10	8	135	2,028
18.	3*	6	7	1	31	19	–	120,000
19.	–	–	–	–	–	28	1	–
20.	–	8	3	4	11	2	3	5,008
21.	1	–	–	–	1	1	–	200
22.	1	2	–	4	3	7	–	3,500
23.	3	–	15	–	11	11	1	3,828
24.	–	–	–	115	10	10	99	5,475
25.	1	–	–	25	30	19	–	13,000

*Includes western and western crossbreeds.

application, which is then passed on to the chairman of the managing committee for action. If approved, a permit is issued allowing the sale to be consummated, although the purchaser is required to sign an agreement that the animals will not be resold and, should the occasion arise, will be returned to the pinjrapole. The cattle involved in these transactions are bred in the institution and are either Gir or graded animals.

Only a few institutions, such as the Bombay, Poona, and Nasik Pinjrapoles, are involved in cattle breeding on a commercial basis, yet this

TABLE 17

SALE OF CATTLE BY THE BOMBAY PINJRAPOLE TO ORGANIZATIONS, 1970–1975

Date	Institution	State	Animals	Cost (Rs).
28-8-70	Cattle Breeding Farm, Kopergaon	Maharashtra	1 bull	1,200
17-12-70	Key Village Center, Sangamner	Maharashtra	1 bull	1,000
25-4-72	Director, Veterinary Services, Indore	M.P.	7 bull calves	13,250
13-6-72	Director, Key Village Scheme, Jaipur	Rajasthan	2 bull calves	3,000
24-3-73	Dept. of Animal Husbandry, M. Phule Krishi Vidyapeth, Rahuri	Maharashtra	44 heifers	15,800
28-3-73	Dept. of Animal Husbandry, M. Phule Krishi Vidyapeth, Rahuri	Maharashtra	36 heifers	12,600
30-6-73	Dept. of Veterinary Services, Indore	M.P.	3 bulls	58,000
29-3-74	District Animal Husbandry Office, Osmanabad	Maharashtra	45 heifers	21,600
14-4-75	Kandivali Cattle Breeding Farm, Bombay	Maharashtra	26 heifers	25,000

(Data provided by the Secretary, The Bombay Pinjrapole.)

does represent an added dimension to the economic activities of animal homes which has emerged over the last few decades.

Economic Functions and Ecologic Mechanisms

What, then, can be said of the economic functions of goshalas and pinjrapoles in India today? It is apparent that, despite considerable variations in the scope of their activities, animal homes do operate as economic institutions, reflecting not only traditional patterns of resource utilization in the Indian subcontinent but also more recent trends in resource development. Their economic activities center around cattle

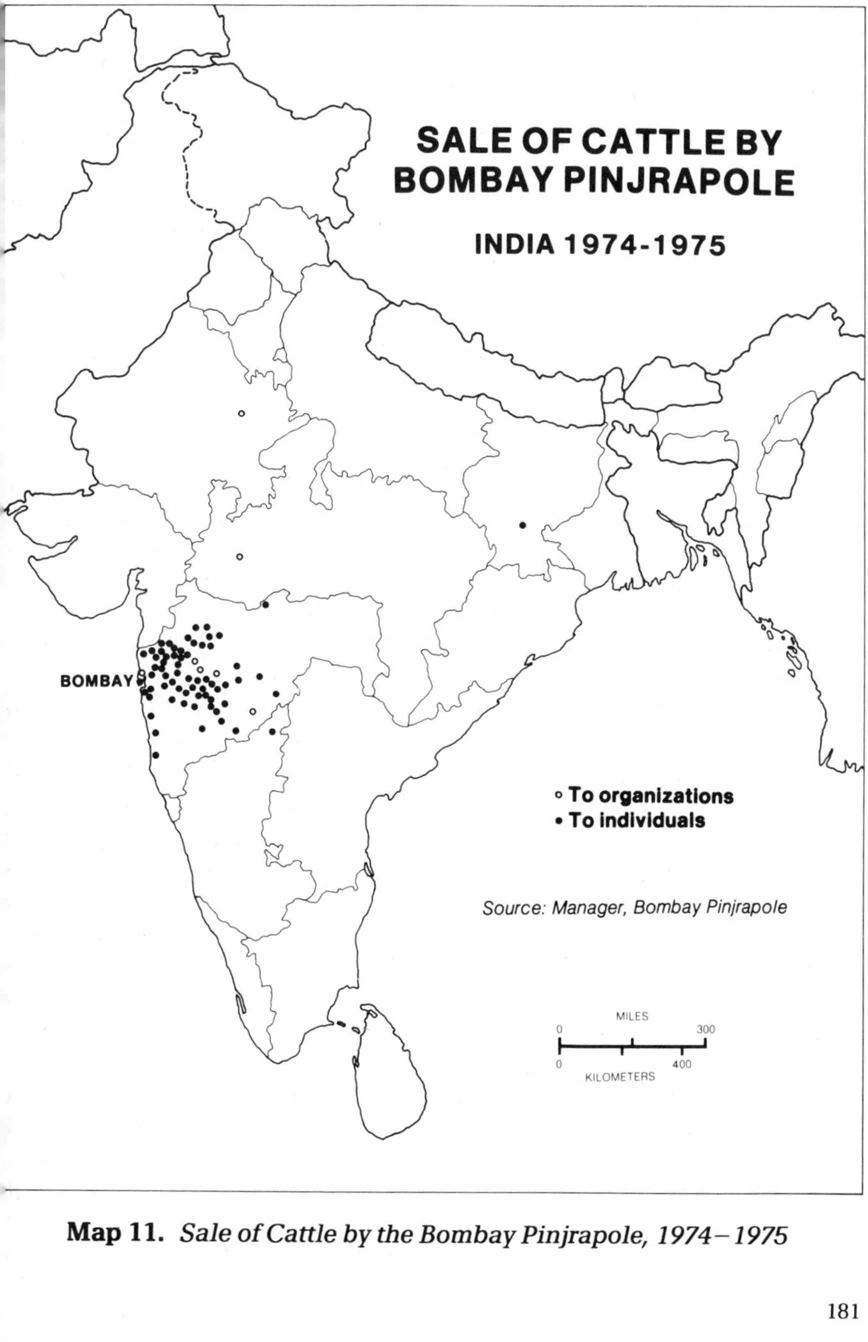

Map 11. *Sale of Cattle by the Bombay Pinjrapole, 1974–1975*

and even in the most traditional institutions catering only to nonproductive animals, dung, hides, and carcasses are by-products of their cattle-protection role. Returns from these sources are minimal, however, and the main economic activities of animal homes center on milk production and cattle breeding. Not all goshalas and pinjrapoles participate in these commercial undertakings. Table 16, for example, shows considerable variations in the percentage of productive cattle in institutions, with some (such as the Kapadvanj and Nadiad Pinjrapoles) being devoted entirely to useless animals. In Patan, on the other hand, over 80 percent of the cattle are useful, and several other institutions maintain sizable dairy herds. These animal homes make positive contributions to the Indian economy in terms of milk supplies and as a source of milk cows and working oxen. This economic role has received recognition in the Indian government's attempts to incorporate goshalas and pinjrapoles into national economic planning as a means of developing the cattle resources of the country.

If the economic has superseded the religious in many institutions today, further dimensions of their activities might be termed ecological rather than economic in that they salvage and return to productivity cattle that have been abandoned by their owners. This function is especially apparent in the large urban areas, where numbers of unwanted animals are often increased by traditional dairying practices. In Calcutta, for example, there occurs a considerable in-migration of milking zebu, estimated by the State Department of Animal Husbandry to number some 500 to 600 animals per day from as far away as the Punjab. Like most conurbations in India, the Greater Calcutta area, with a population of over twelve million, experiences chronic milk shortages and provides a ready market for these animals.[12] However, the cost of keeping cows through their dry period in urban areas without any immediate returns is so great that many animals are disposed of after their first lactation, usually being sold to the slaughterhouse.[13,14] Some of these animals are purchased by Hindus and Jains and are placed in the Calcutta Pinjrapole, while the institution itself occasionally purchases animals culled from government cattle farms and destined for the slaughterhouse.[15] The actual numbers and potential economic contributions of these animals, however, are limited by the availability of funds.

Of much greater significance is the role of goshalas and pinjrapoles in protecting cattle during times of natural calamities. As previously noted, this has been put forward as the reason for the founding of some institutions and is well illustrated by events in Gujarat during the famine and drought in 1972–1974. The activities of the pinjrapole in Mandal, a small town of 10,000 people located some eighty kilometers northwest of

Ahmedabad, were typical of many institutions at this time. In March 1975, the animal population of the Mandal Mahajan Pinjrapole included 50 goats, 500 buffalo, and some 1,500 cattle. Many of these animals were potentially productive milking cows and draught oxen placed in the institution only because of their owners' inability to feed them under the prevailing drought conditions.

The pinjrapole follows the practice of releasing useful animals free of charge to any individual who can show that he has need of cattle and is capable of maintaining them properly. The aspiring recipient submits an application to the pinjrapole for the number of animals he requires, accompanied by a certificate from his *sarpañc* (head of the panchayat) attesting to his need for them and his capability of feeding them, and by references from two persons who stand as guarantors of the arrangement. The person requesting the animals also signs an agreement that they will not be sent to a slaughterhouse, will not be sold to any third party, and will be returned to Mandal if they can no longer be maintained. The application is then submitted to the managing committee of the pinjrapole which approves or denies the request. Between 1 April 1974 and 1 April 1975, a total of 405 animals, buffalo as well as cattle, were donated to 127 applicants. Most of the requests were for one or two animals, but on one occasion 275 cattle were given to a group of Ādivasis from Kavitha village in Baroda District. Recipients came from all castes and communities, including tribals, Rabari pastoralists, Moslems, and Hindu agriculturalists. Most of the people receiving animals were from nearby districts, though some came from as far as Kutch, over one hundred fifty kilometers away (map 12).

We see here in action, therefore, an ecological mechanism that serves to protect cattle resources at times of climatic adversity. The poorer agriculturalists, marginal farmers, and even pastoralists who choose not to migrate to better-watered locales can place in the asylums animals that would otherwise perish in periods of drought and fodder scarcities. Not all of these animals survive, raising questions as to the effectiveness of this mechanism, yet many are kept alive until conditions return to normal and are then placed back into agricultural use.

The financial structure peculiar to animal homes aids in this cattle-protection role. The various lags supporting the institutions are imposed mostly on sales of grains and other agricultural produce, on manufactured goods, such as cotton textiles, based on local resources, and on the transportation of commodities—all of which are crude measures of the health of the local economy. These informal taxes withdraw a small percentage of the money supply changing hands in business transactions relating essentially to locally based industries, making it available

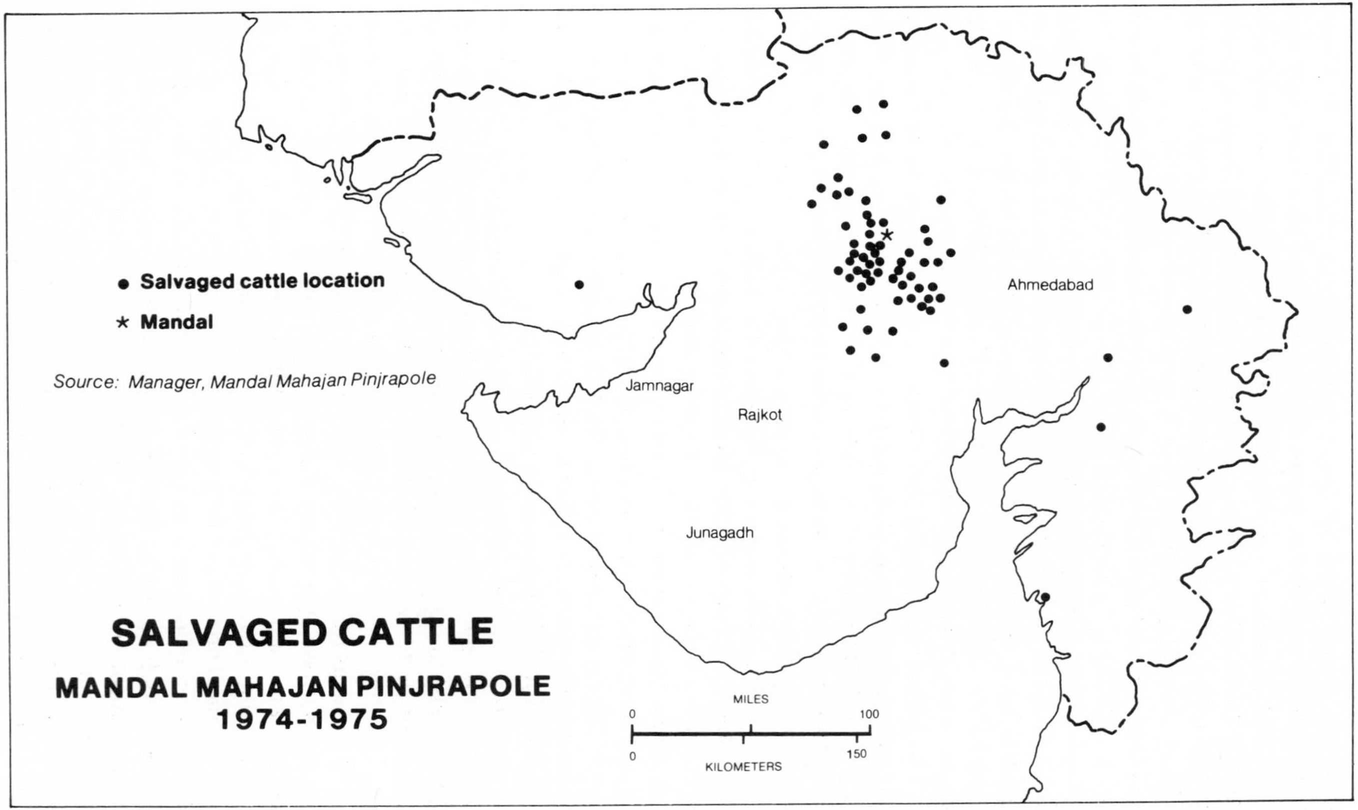

Map 12. *Salvaged Cattle from the Mandal Mahajan Pinjrapole, 1974–1975*

for the protection of cattle, on which many of these industries ultimately depend.

In Gujarat, moreover, goshalas and pinjrapoles serve as a means of channeling funds from the wealthy vania communities of the major cities into agricultural regions where cattle play a major role in the economy and which are most susceptible to the damaging effects of drought. This process, whereby urban wealth is recycled back into the system to protect the cattle resources on which much of this wealth is built, depends on the traditional ties between the urban vania communities and their native towns as well as on their special regard for values such as ahiṃsā and reverence for the cow. No doubt this protective mechanism functions elsewhere in India, yet nowhere else is it as effective as in Gujarat and western India.

It can be seen, therefore, that animal homes engage in a variety of economic activities. Some institutions make positive contributions to the agricultural economy through commercial milk production and cattle breeding. Where goshala lands are cultivated, the main efforts are directed toward fodder production, thus reducing the supplies needed to be withdrawn from the open market and even making agricultural produce available for sale. Even institutions not engaging in commercial activities do produce dung, hides, and carcasses, thereby contributing, however minimally, to the local resource base. Yet there are dysfunctions in goshala and pinjrapole activities. Much dung is lost; hides and carcasses, if used at all, are of extremely low quality; valuable agricultural land is withdrawn from other uses to feed nonproductive animals; and the preserving of useful cattle during periods of drought and fodder scarcity has to be balanced against the consumption of fodder by useless animals at other times. Both the positive and the negative nature of these activities, however, are derived from essentially the same source—the institutions' prime concern for the protection of cattle.

X

Animal Homes in National Planning and Economic Development

Economic functions of goshalas and pinjrapoles have assumed increasing importance during the last three decades in the broader context of national planning and economic development. After gaining independence from British rule in 1947, India embarked on a policy of planned and directed growth designed to transform an underdeveloped, colonial economy into one befitting the second most populous nation in the world.

For Indian planners, the basic problem of development was one of utilizing more effectively resources available to the country.[1] Animal homes, as institutions concerned with cattle welfare, were assigned a role in the development of the country's vast cattle resources. According to the 1951 livestock census, there were some 150 million cattle and 43 million buffalo in India, contributing an estimated Rs. 10 billion to the gross national income. Some 10 percent of the total cattle population were unserviceable or unproductive, yet these were consuming fodder supplies sufficent to meet only 78 percent of the country's requirements.[2] The situation at the beginning of the planning period is summarized in the words of the first Five Year Plan:

> . . . the available feeds cannot adequately sustain the existing bovine population. While there is a deficiency of good milch cows and working bullocks there exists a surplus of useless or inefficient animals; and this surplus, pressing upon the scanty fodder and feed resources, is an obstacle to making good the deficit.[3]

In the minds of the planners, goshalas, pinjrapoles, and the newly instituted gosadans were to contribute to a more effective use of cattle resources in several ways. By segregating diseased and unproductive animals from the rest of the cattle population, they would limit the spread of disease and prevent uncontrolled breeding. By locating unproductive animals in isolated areas with hitherto untapped grazing resources, they would conserve valuable fodder. By confining useless cattle, they would reduce potential damage to crops and the indiscriminate consumption of fodder supplies that could be better utilized by productive animals. Existing institutions could be made more effective by being used as centers for upgrading the quality of cattle, for selective breeding programs, for improved milk production, and for the introduction of modern techniques of animal husbandry.

The first step toward formulating a national policy of development for goshalas and pinjrapoles was taken as early as 1942, when the question of reorganization of the institutions was discussed at a meeting of the Animal Husbandry Wing of the Board of Agriculture and Animal Husbandry. In accordance with the recommendations of the board, a meeting of representatives from goshalas and pinjrapoles was held in New Delhi in 1944 by the Imperial Council of Agricultural Research, with the object of exploring the matter further. It was decided that the institutions should be surveyed to determine their methods of working, their facilities, and their resources. On the basis of this, measures could be implemented for their reorganization and development.

The survey revealed the following facts:

A. Goshalas and pinjrapoles were generally situated either in the vicinity of big towns or in their immediate neighborhood, where feeding and maintenance of cattle was unnecessarily costly.

B. In most of the institutions the tendency was to overstock, with the result that the animals were generally underfed and not properly looked after. This entailed great hardship and suffering for the animals.

C. There was no means of segregating the productive animals from those which were aged, infirm, or otherwise economically useless. Thus, no differentiation was practicable in the feeding and management of good animals from that of unproductive ones, which caused good animals to become worthless after a short time.

D. Land attached to the goshalas was generally insufficient for grazing of cattle or for raising of fodder crops.

E. Promiscuous breeding was practiced as a general rule, and no attention was paid to the selection of bulls for breeding. This resulted in increased numbers of inferior cattle.

F. There were no proper facilities for veterinary aid or treatment of sick animals.

G. A few goshalas had permanent funds managed by duly constituted trusts, but the source of income of most institutions was from dharmada, laga, or goshala cess levied on the sale and purchase of principal commodities in different markets. Some goshalas lacked the patronage of any business community and depended entirely on public charities and donations. The income of these institutions was spasmodic, irregular, and very uncertain, which was a great handicap in envisaging permanent improvement in their management. Moreover, a substantial part of the money deducted as dharmada was not rightly spent, and its actual transfer to the goshala account rested entirely on the discretion of the persons who collected the funds.

H. The institutions were generally managed by untrained staff, so there existed no planned policy according to which the management, breeding, and feeding of different types of animals could be carried out.[4]

One result of these findings was the creation by the Indian government of the post of Cattle Utilization Adviser, whose task it was to direct the reorganization and development of goshalas and pinjrapoles in the country.

The immediate postwar years saw increasing government involvement in the development of the animal homes, mainly at the provincial level but under advisement from the national capital. By 1948 Goshala Development Officers with the duty of rendering all possible assistance necessary for the development of the institutions had been appointed in seven states and provinces.[5] Eight provinces had Goshala Development Schemes in operation, providing financial and technical assistance to selected institutions, while such schemes were under consideration by other provincial governments.[6]

This period also saw the formation of the Provincial Federations of Goshalas and Pinjrapoles to coordinate and standardize the activities of institutions within the states and provinces of India. These were nonofficial organizations established and run by members of the goshalas themselves, and were conceived of as an unofficial link between the institutions and the government. Membership by goshalas was purely voluntary, but affiliated institutions were expected to conform to federation rules and requirements concerning organization, the maintenance of accounts and general operating procedures. By early 1949 federations of goshalas and pinjrapoles were functioning in eleven provinces and states,[7] many of which subsequently enacted legislation aimed at providing better management, control, and development of goshalas and pinjrapoles within their territories. In Rajasthan, for example, the Rajasthan Gaushala Act of 1960 requires that every goshala register with the State Registrar of Gaushalas and it sets out rules and regulations for the operating of the institutions.[8]

Though these various schemes were implemented at the provincial level, much of the information gathering, planning, and drafting of model programs was carried out in New Delhi. As the activities of the Cattle Utilization Adviser's section in the Department of Agriculture expanded, so did its staff, with the posts of Assistant Cattle Utilization Adviser and Deputy Adviser being created in 1946 and 1948, respectively. In addition, in November 1948, a Central Goshala Development Board was constituted specifically for the purpose of overseeing goshala development in the country. It was later replaced by the Central Council of Gosamvardhana.

Central Council of Gosamvardhana

In 1952 the Indian government decided to establish an all-India organization to supervise cattle-development activities throughout the country, setting up for this purpose the Central Council of Gosamvardhana, a body composed of government officials, agricultural and animal husbandry experts, representatives of the State Federations of Goshalas and Pinjrapoles, and of State Councils of Gosamvardhana.[9] The council was to replace the Central Goshala Development Board, with its more limited objectives, and to concern itself with all aspects of developmental activities, ranging from implementing milk and cattle schemes to publishing promotional literature.

The central council continued the work of its predecessor in sponsoring Goshala Development Schemes and in providing technical and financial assistance for the improvement of the institutions. Its recommendations on gosadans and goshalas were accepted by the planning commission and were incorporated into the Five Year Plans. During 1954 and 1955, a comprehensive survey of goshalas and pinjrapoles throughout India was undertaken as a basis for drawing up plans for future development. The results were subsequently published.[10] Studies were made of various schemes for the salvage of dry milch cattle from urban areas and of ways to increase fodder supplies. Gosamvardhana seminars were held to discuss the work of the central council.[11] The workings of State Federations of Goshalas and Pinjrapoles and State Councils of

Gosamvardhana were reviewed, and measures were suggested t improve their functioning. The activities of the council included th publication of a monthly journal, *Gosamvardhana*, which served as forum for debate and discussion by all those interested in cattle devel opment in India.

The effectiveness of the Central Council of Gosamvardhana as a na tional coordinating body, however, was severely restricted by its advisor nature. With no statutory powers, the council could only offer recom mendations with no means of ensuring that they would be implemente or that there would be the necessary fiscal powers to fund them.[12] More over, with animal husbandry being in the hands of state governments directives from the central government were not always implemented Concern that the advisory character of the Central Council of Gosamvar dhana would reduce its potential contribution to the preservation an development of India's cattle wealth had been expressed at its very in ception, and in 1960 steps were taken to expand the scope of its activi ties. The reconstitution of the council at this time, however, did no extend to granting it statutory powers.

The scope of the newly reorganized Central Council of Gosamvar dhana, as set out in the Ministry of Food and Agriculture (Department o Agriculture) Resolution No. 7-16/59-L.D. dated 6 December 1960 wer as follows:

> 2. The functions of the Council shall be:
>
> (a) To organise, implement and coordinate activities relating to the preserv tion and development of cattle, and generally, to administer the scheme relating there-to for the greater production of milk and increase of draugh power.
>
> (b) To organise, and coordinate the State Councils of Gosamvardhana, Fe erations of Goshalas and Pinjrapoles on matters relating to the develop ment of cattle wealth and establishment and development of Goshala on proper lines.
>
> (c) To establish Key-village Centres for the breeding of cattle on scientifi lines and the starting of Gosadans for bovine cattle and to diffuse usefu scientific knowledge on animal husbandry throughout India.
>
> (d) To sponsor schemes relating to the increased production of feeds an fodder, improvement and development of pastures and grazing areas salvage of dry cattle, rearing of calves, rounding up of wild and stra cattle, running of training centres and other allied subjects.
>
> (e) To take steps for the prevention and eradication of infectious and conta gious diseases affecting the life and health of bovine cattle and also tak adequate steps for preservation of cattle in times of famine and othe emergency situations.
>
> (f) To review from time to time the progress of schemes relating to preser

vation and development of cattle in the light of the coordinated programme and policy laid down for the country and to consider such additions and alterations to the programme as may be found necessary in the light of experience gained.

(g) To take such steps as may be necessary to implement provisions of the Constitution relating to the Organisation of Animal Husbandry as expressed in Article 48.

(h) To collect statistics in respect of the cattle population of the country, number of Goshalas and Pinjrapoles and other matters referred to above.

(i) To carry on propaganda for the promotion of the objectives here-in-before mentioned.

(j) To make such other measures for Gosamvardhana including those mentioned herein as may be considered necessary from time to time.

(k) To advise the Central and State Governments concerned on any point referred to it by them.

(l) To employ such staff as may be necessary for the proper performance of any or all of these functions.

(m) To adopt and undertake any other measures or performing any other duties which may be required by the Government of India to adopt or perform or which the Society may consider necessary or advisable in order to carry out the purposes for which it is constituted.

(n) For the purposes of the Society to draw and accept and make and endorse, discount and negotiate Government of India and other promissory notes, bills of exchange, cheques or other negotiable instruments.

(o) To permit its funds to be held by the Government of India.

(p) To purchase, take on lease, accept as a gift or otherwise acquire any land or building or works wherever situated in India which may be necessary or convenient for the Society and to construct or alter and maintain any such buildings or works.

(q) To issue appeals for funds in furtherance of the objects of the Society, to receive gifts and undertake the management of any endowment trust funds or donations not inconsistent with the objects of the Society.

(r) To sell, lease, mortgage or exchange and otherwise transfer all or any portion of the properties of the Society.

(s) To establish and maintain research and reference libraries and reading rooms.

(t) To offer prizes and grant scholarships in furtherance of the objects of the Society and to finance the examiners and researchers.

(u) To establish a provident fund for the benefit of employees.

(v) To give grants to further the objects of the Society.

(w) To do all such other things either alone or in conjunction with others as the Society may consider necessary incidental or conducive to the preservation and development of the cattle for the greater production of milk and increase draught power.

Despite such a sweeping mandate, the Central Council of Gosamvardhana still lacked any real power to implement its recommendations. It continued its work for several years, making valuable contributions to the preservation and development of India's cattle wealth, but it was finally disbanded in 1969 with its responsibilities being assumed by the relevant government departments. Two aspects of the council's activities merit further discussion, however, since they both relate to animal homes in India's Five Year Plans.

The Gosadan Scheme

The gosadan was seen by the authors of India's first Five Year Plan as an answer to the pressing problems of unproductive cattle and their adverse effect on the economy. The removal of useless cattle was accorded a high priority in the plan for livestock development, which called for the establishment of 160 of the institutions, complete with tanneries, over the period from 1951 to 1956. Each gosadan was to house some 2,000 cattle, so that by the end of the five-year period 320,000 animals would be maintained in the institutions. Preference was to be given, in removing cattle, to surplus animals from areas where the key-village scheme for cattle development was in operation. The total cost of the Gosadan Scheme during the first Five Year Plan was set at Rs. 9.7 million, amounting to some $2 million in 1951.[13]

The scheme did not meet with much success. The initial funding was reduced to Rs. 1.5 million and the proposed target of 160 institutions was cut drastically to 40. By 1955 only 24 centers had been sanctioned and 1[illegible] actually set up, housing 5,293 cattle instead of their planned capacity of 34,000.[14] Public support for the gosadan was lacking, and a variety of other factors contributed to the failure of the scheme. Difficulties were experienced in locating suitable land, and transportation facilities were found to be inadequate. State governments were unable to meet their share of the expenditures,[15] and the absence of legislative measures for the compulsory removal of unwanted and stray cattle reduced the effectiveness of the scheme. Indeed, the value of the entire program was questioned by the Committee on the Prevention of Slaughter of Cattle in India, which stated,

> In view of this [huge recurring expenditure] and in view of the indifferent responses from the states in setting up Gosadans, the Committee are of the view that the Gosadan scheme is not likely to offer a solution for the problem of useless cattle and that it would be far more desirable to utilize the limited resources of the country to increase the efficiency of the useful cattle.[16]

Similar reservations concerning the progress of the Gosadan Scheme were expressed in the planning commission's review of the first plan, noting that only twenty-five of the institutions had been set up during the period 1951–1956, handling a total of some 22,000 cattle—a far cry from the one-third million animals envisaged at the outset of the plan. Recommendations for the improvement of the scheme included provision of free transport for cattle, use of existing goshalas as collection centers for old and useless cattle, and financial assistance for the establishment of gosadans by private organizations.[17] The second Five Year Plan, also, concluded that gosadans alone did not hold the solution to the problem of surplus uneconomic cattle and that

> . . . in defining the scope of bans of the slaughter of cattle States should take a realistic view of the fodder resources available and the extent to which they can get the cooperation of voluntary organizations to bear the main responsibility for maintaining unserviceable and unproductive cattle with a measure of assistance from the Government and general support from the people.[18]

Despite its failings, the Gosadan Scheme was continued, though on a much less ambitious scale, during the subsequent plan periods. Thirty-six institutions were established during the second plan, and a further twenty-three during the third Five Year Plan, which also provided for the establishment of collection centers, improved transport facilities, construction of tanneries at gosadans, and increased subsidies to private institutions. Certain gosadans were taken over from the state governments by the Central Council of Gosamvardhana, which showed that, if they were properly managed, the institutions could be economically sound.[19]

Although attempts were made to increase public involvement through the establishment of private gosadans, the experiment was not always successful, as is illustrated by the experience of one landowner in eastern Uttar Pradesh. According to the informant, he was approached by the state about the possibility of setting up a private gosadan, which was established in 1959. The institution served five neighboring districts of the state, with animals being collected through the government's stockman centers. Villagers would be informed of the days on which collections were to be made and would bring their useless cattle to the center, whence they would be transported by truck to the gosadan. A government subsidy of Rs. 1.50 per animal per month was paid for the maintenance of cattle in the gosadan, which had 700 acres of grazing land, adequate for some 400 head of cattle. A small tannery was set up and equipped by the government and was staffed by a government-appointed master flayer who was to teach the local villagers good flaying techniques. A small cooperative society was also formed for the utilization of carcasses.

In theory, the institution conformed to the model set out for gosadans, but in practice it worked somewhat differently. It rarely functioned at full capacity, despite serving five districts. The subsidy was inadequate to pay for the salaries of herdsmen and incidentals such as medicines, and often was not received at all. The flayer, who oddly enough was a Thakur by caste, did no flaying, so this had to be done by the owner's own men. Difficulties were experienced in selling the bone meal and meat meal processed by the tannery, which eventually shut down, as did the cooperative. The hides are still purchased by contractors, but the local harijans, acting under a directive from leaders not to flay or even bury the cattle of caste Hindus, will not touch the carcasses. Though in years past they would eat carrion meat, this practice has long since been abandoned. The raw bones of dead animals are collected by Doms,[20] who sell them to contractors for processing in the larger fertilizer factories. The gosadan accounts for the year 1970–1971 show that income from hides and bones was Rs. 3,786 while expenditures were Rs. 9,151. Although gosadans can be self-supporting, many of the smaller institutions, especially the private ones, are beset by difficulties that make them inefficient and uneconomic to run.

The realization that gosadans did not provide an effective solution to the problem of unproductive cattle was reflected in their declining role in the later plans. By 1966, the end of the third plan period, eighty-four institutions had been set up, a little over half the number initially proposed for the first Five Year Plan alone. Despite the feeling of the Gosamvardhana Seminar at Bombay that at least 200 more of the institutions should be set up during the fourth plan period,[21] the gosadans received scarce attention in the annual plans introduced in 1966 and in the fourth and fifth plans covering the years 1969–1979. The direction of planning and allocation of resources placed much greater emphasis on intensive cattle development than on the removal and preservation of unproductive animals. Some new gosadans were begun by the states during this period, but a general lack of public support, scarcity of funds, and a low level of priority at the state level have reduced the Gosadan Scheme to a mere shadow of that originally envisaged by Indian planners. Today, perhaps one hundred of the institutions are to be found in India, offering shelter to sick, aged, and infirm cattle—monuments to the values of the past but also living reminders of the economic realities of the present.

The Goshala Development Scheme

Whereas the Gosadan Scheme was designed to meet the problem of nonproductive cattle, the goshalas and pinjrapoles were seen as a means of improving the quality of India's cattle resources and as centers of cattle and milk production. Although the original survey of the institutions undertaken in 1944 revealed their shortcomings, it also revealed their potential. It was estimated that 20 percent of the animals maintained in the goshalas and pinjrapoles were of good dairy types and that a further 20 percent were fit for breeding. With proper organization and management and adequate funding, they were thought capable of producing over one million maunds (37 million kg.) of milk annually, some 25,000 stud bulls, approximately the same number of work bullocks, and some 50,000 improved-quality female calves.[22]

Various Goshala Development Schemes were initiated in the states and provinces as early as 1945 but were not implemented at the national level until the second Five Year Plan. In this, it was proposed to select 350 institutions for development as centers of livestock improvement, with Rs. 10 million (ca. $3.5 million) being set aside to finance the scheme during the plan period.[23] Provisions for technical and financial assistance included:

1. Supply of ten purebred cows and a good stud bull by the government, subject to the condition that the institution concerned would also acquire a similar number of approved cows at its own cost.
2. A nonrecurring grant of Rs. 5,000 for the improvement of existing buildings, development of land, purchase of equipment and other necessities.
3. An annual recurring grant of Rs. 2,000 to meet the increased expenditure on the maintenance of the productive herd.
4. Subsidy at the rate of Rs. 10 per calf per month for rearing selected purebred calves available at each of these goshalas.
5. Extension of financial assistance to the State Federations of Goshalas and Pinjrapoles to enable them to run regular offices and to coordinate effectively the goshala development activities in the states.
6. Provision of goshala development staff.[24]

In addition, free veterinary aid was to be provided, and the Central Council of Gosamvardhana was to maintain an inspection and advisory staff in addition to running training centers for goshala workers.[25] Thus, the institutions chosen for development would be provided with a nucleus herd of high-quality milch cows, would be assured of a steady income, and would receive the necessary technical guidance and assistance to maintain the herd in good condition.

As might be expected with any new undertaking, certain difficulties arose with the scheme. It was felt that goshala managements sometimes lacked enthusiasm for developmental work, their halfhearted efforts detracting from the effectiveness of the program. In one instance, the local Marwari community, which supported an institution selected for development, rejected the scheme and refused to meet any of the expenditures required by it. Some goshalas and pinjrapoles, not being exempted from various land-ceiling and tenancy acts, experienced problems in maintaining adequate grazing for their animals. In urban areas, the high cost of stall-feeding animals was such that the annual maintenance grant was insufficient, with a resulting decline in the condition and productivity of quality cattle. Deteriorating economic conditions slowed down the implementation of the Goshala Development Scheme in many states.[26]

Despite these handicaps, 255 goshalas and pinjrapoles were taken up for development during the second Five Year Plan, exceeding the revised plan target of 246 institutions. Up to 31 March 1961, 212 breeding bulls and 3,978 cows had been supplied to institutions in the Goshala Development Scheme, and the rearing of 1,244 bull calves had been subsidized. During the year 1960–1961, participating goshalas produced 5,028,193 kg. of milk,[27] and the breeding program also produced results. In Gujarat, for example, the goshalas made available to the State Department of Animal Husbandry 725 bulls and 75 cows. A total of 344 bulls, 860 cows and heifers, and 606 bullocks were sold to private institutions and farmers for dairying and agricultural purposes.[28]

The results of the Goshala Development Scheme during the second plan were encouraging, and it was continued during the third with plans for the development of an additional 168 institutions. The revised scheme made provisions for the subsidy of up to 50 percent of the salary for trained managers in goshalas as well as loans for the improvement of irrigation facilities and the expansion of fodder-producing capabilities. Recommendations were also made by the Central Council of Gosamvardhana that financial assistance should be continued beyond the usual five years to maintain continuity in development, that the Goshala Scheme should be designated a pattern scheme so that it might receive higher priority, and that selected institutions should be used for agricultural extension work.[29] These were taken under consideration by the government but were never implemented.

By 1966, the end of the third Five Year Plan, an additional 122 goshalas had been taken up for development, but during the last decade the role of the institutions in national planning had diminished considerably. Economic conditions forced the adoption of a series of annual plans between 1966 and 1969, and although the scheme was continued through the fourth plan (1969–1974), it received no mention in the draft of the Fifth Year Plan.[30] Goshala Development Schemes continued to operate intermittently in states such as Haryana, Rajasthan, and Gujarat,[31] but they never again achieved the success of the late 1950s.

Efforts during the last three decades to incorporate animal homes into national economic planning have met with mixed success. The gosadan, unless properly managed and equipped, has been found to be uneconomic, and it could never be implemented on the scale necessary to affect the more than eleven million useless cattle in India. The Goshala Development Scheme, however, has succeeded in increasing milk production; it has succeeded in improving the caliber of goshala herds; and it has succeeded in producing numbers of good-quality cattle for the agricultural economy. The idea of using goshalas to further economic development was attractive, and under different circumstances it might well have achieved more than a limited success. However, the lack of a central body with statutory powers to implement the scheme, low priority levels, scarcity of funds and periods of difficult economic conditions have all served to detract from the contributions of animal homes to livestock development and to the role envisaged for them in India's national economic planning.

XI

Animal Homes and the Sacred Cow of India

The animal homes of India are unique cultural institutions well worthy of attention in their own right, but as elements of the Indian cattle complex their study also raises issues of broader significance concerning man and cattle in India. Is the sacred cow of India a myth or a reality? Are the "irrational, non-economic and exotic aspects of the Indian cattle complex . . . greatly overemphasized at the expense of rational, economic and mundane interpretations," as Harris suggests? Do the various traits associated with the sacred cow concept stem from utilitarian rather than ideological considerations? Can the alleged mismanagement of India's cattle resources be interpreted as reflecting anything other than the negative influence of strategies of action embracing concepts such as ahiṃsā and other Indian religious beliefs? These are some of the questions concerning the sacred cow concept that have been debated with much vigor in the scholarly literature during the last decade, and the same questions can be raised about goshalas and pinjrapoles. Are these homes *only* ecological mechanisms for preserving cattle, and are their social and religious characteristics *merely* peripheral and to be discounted, or are they phenomena reflecting complex social, ideological, and environmental forces that have shaped attitudes and behavior toward animals in India over the centuries? Are they positive-functioned and adaptive, or dysfunctional, or both? Is their presence detrimental to animal husbandry in India? What, if anything, do they contribute to the agricultural economy of the country? How do the institutions fit into the various interpretations of the functioning of the sacred cow doctrine in India today?

This study has shown that animal homes in India are essentially religious institutions. They express basic concepts that are overwhelmingly religious in nature, and they have assumed a religious character in modern Indian society. Even where their animal-protection activities extend to cattle, a consideration of their economic functions in the broader context of the cattle-keeping economy leads to the same conclusion. The preservation of useless cattle in goshalas and pinjrapoles, for example, can in no way be justified in purely economic terms. Some 29.95 percent of the cattle in animal homes are nonproductive, and thus with an estimated 580,000 animals in the institutions, close to 174,000 useless cattle are being maintained for minimal or no return. Their only economic potential lies in the residual value of their hides and in their dung, and it has already been shown that methods of carcass disposal commonly employed by goshalas and pinjrapoles inhibit full exploitation of this resource. The only possible basis for maintaining these otherwise nonproductive cattle remains their dung, and their continued survival as mere dung-producing machines makes little economic sense. The allocation to useful animals of fodder now consumed by them would not necessarily reduce the amounts of dung available and would yield additional benefits through the increased productivity of the working animals.

Furthermore, not only are the potential returns from useless cattle minimal, but costs incurred in their upkeep far exceed any income derived from their dung, hides, and carcasses. Table 18 shows the breakdown of costs relating to the upkeep of cattle at the Rajkot Mahajan Pinjrapole during 1958. All figures are expressed as percentages of the year's total outlay of Rs. 36,897. Of this sum, 76.3 percent was spent on the purchase of feed and the remainder on water costs, workers' salaries, and medicines for the animals. Cash returns from the sale of dung and carcasses amounted to 20.6 percent of expenditures, while the sale of live animals brought in an additional 4.9 percent. In other words, for every rupee earned from its animals, the pinjrapole spent four on their upkeep. These figures, moreover, do not include additional inputs such as administrative overheads and the value of feed donated as dharmada.

As might be expected, the economics of supporting useless animals vary considerably depending on the practices followed in the institutions concerned, but all animal homes that maintain useless herds operate at a deficit. In the Ahmedabad Pinjrapole, for instance, where carcasses are removed by Chamars rather than sold as at Rajkot, comparative figures

TABLE 18

CATTLE-RELATED COSTS AND INCOME, RAJKOT MAHAJAN PINJRAPOLE, 1958

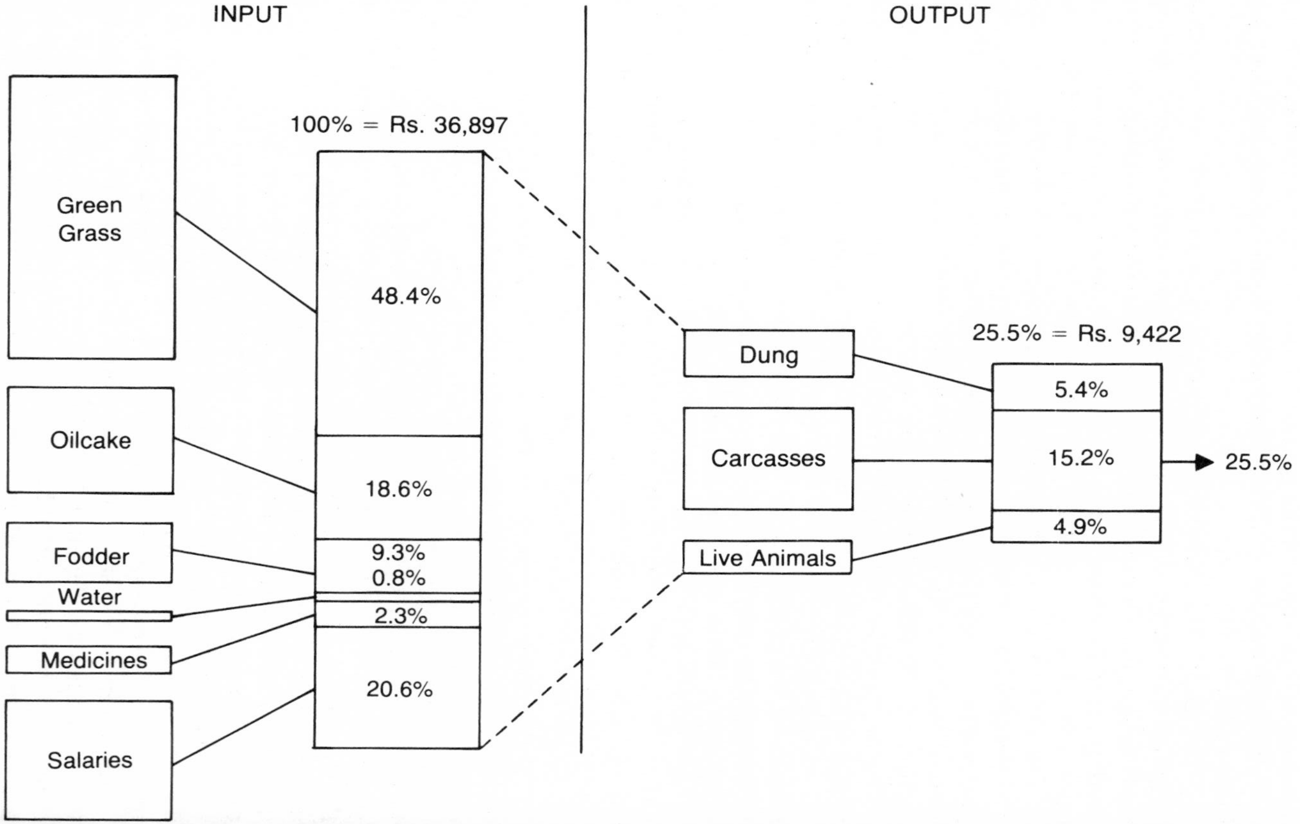

(though including all the herd animals and not just cattle alone) show fifty rupees spent for every one rupee earned. Even in institutions such as the Vrindavan and Dibrugarh Goshalas, which support milk herds, the maintenance costs of nonproductive animals far exceed the profits generated by the useful ones. These institutions spend Rs. 2-80 and 1-90, respectively, for every rupee earned through the sale of milk, calves, and dung.

The preservation of useless cattle in animal homes is, therefore, accomplished at a considerable financial loss, the deficit being made up by dharmada, rent, and other funds available to institutions. Not only is this practice uneconomic, it also raises questions concerning the efficiency of resource utilization. Most institutions included in the sample survey estimate the daily consumption of fodder to be 10 kg. per animal, and thus useless cattle in goshalas and pinjrapoles consume over one-and-three-quarter million kilograms of fodder every day in a country facing chronic shortages of fodder for its cattle population! The economically harmful effects of animal homes goes beyond this misallocation of resources, since useless animals in the institutions compete directly with useful cattle for available supplies in times of scarcity. Goshalas and pinjrapoles, with their greater purchasing power, can effectively outbid local farmers for fodder during periods of shortages, so that useless and unhealthy cattle that would normally be the first to succumb under such conditions survive at the expense of useful animals.

There are, however, some positive aspects to the cattle-protection activities of goshalas and pinjrapoles. As seen in Gujarat during the 1972–1974 drought, animal homes, with their greater financial resources, were able to truck in fodder supplies from other parts of the country and so preserve useful animals placed there by farmers who could no longer maintain them. Yet, while many of these animals were later returned to the agricultural sector of the economy, many also perished, raising doubts as to the efficiency of this protective mechanism. Mortality rates at the Ahmedabad Pinjrapole, for instance, were as high as 87 percent in 1973. There is no question that only a small percentage of the useful animals placed in institutions might ultimately be returned to a productive life. Other things being equal, however, if animal homes did not exist, even these useful animals would perish. Furthermore, the basic objective of animal homes is the preservation of animal life in general and not the preservation of useful cattle alone. This protection mechanism, however inefficient, must be seen as secondary to the main activities of goshalas and pinjrapoles and not as their primary purpose.

There are thus both negative and positive aspects to the activities of goshalas and pinjrapoles; the negative are essentially related to religious functions and the positive to economic activities, but the two often encroach on each other. It is, moreover, difficult to balance the positive against the negative. How, for instance, can the maintaining of useless animals in goshalas and pinjrapoles in normal conditions be equated with the preservation of useful cattle in times of extreme scarcity? Is it possible to justify the animal-protection activities of the institutions in terms of the numbers of cattle salvaged and returned to a productive life?

From a strictly economic point of view there is no doubt that the preservation of useless cattle in goshalas and pinjrapoles is detrimental to the cattle-keeping economy of India and does represent a mismanagement of resources. Considerable financial losses are incurred in supporting these animals; and the consumption of fodder by useless cattle, the use of agricultural land to produce fodder for these animals, and minimal exploitation of their dung, hides, and carcasses all detract from the most efficient use of available resources. Yet there remains the question of alternatives. What should be done with the useless cattle that find their way into animal homes? The rational, economic solution, from the point of view of resource utilization, would be slaughter—but if Indian animal husbandry were solely governed by rational, economic practices the problem of useless cattle would not even arise. This is not to imply that cattle keeping in India functions in an essentially irrational, uneconomic manner, but rather that it operates within a cultural context that places emphasis on values other than efficiency alone. The recent popular agitation in India for a total, nationwide ban on cow slaughter is but an example of the extent to which cultural attitudes may be out of phase with the dictates of economic common sense. Despite their economic functions, goshalas and pinjrapoles are outgrowths of a culture that places great value on the protection and preservation of animal life, and especially of the cow. In sheltering injured animals and useless cattle, the institutions provide an outlet for expressing these cultural values, an outlet which may make little economic sense, but one which truly reflects attitudes current in contemporary Indian society.

Animal homes in India cannot, therefore, be evaluated strictly in economic or utilitarian terms. How can a price be placed on the personal satisfaction derived by the devout Hindu from the performance of religious observances? How can apparently irrational practices, such as the preservation of insect life, be weighed against the spiritual rewards that patrons of animal homes believe themselves to have accrued as a result of their actions? How can the inefficient and nonproductive use of resources in goshalas and pinjrapoles be balanced against the social cohesion and the community identity that animal homes bring to Marwaris in Assam? The answer is simple—they cannot. Goshalas and pinjrapoles must be seen for what they are: complex social institutions that reflect the interaction of forces and processes that have been at work in the Indian subcontinent for hundreds of years. They fulfill social and religious functions as well as economic ones, and this is a point that is either missed or ignored by many of those who seek to interpret the functioning of the sacred cow concept solely in terms of economic materialism or cultural ecology. Once committed to the primacy of economic or ecological factors, then all facets of behavior related to the sacred cow concept, such as beef avoidance, reluctance to kill cattle, belief in ahiṃsā, and support of animal homes must ultimately be explained in economic or ecological terms.

As far as animal homes are concerned, such interpretations can be challenged on religious, historical, and ecological grounds. Is the presence of useless cattle in goshalas and pinjrapoles to be explained as the result of undesirable, though "positive-functioned" and "adaptive" processes at work in the cattle economy of India? It seems to this author more likely to be a reflection of beliefs, attitudes, and behavior originating in a cultural tradition in which religion, economy, and society are inseparably intertwined—an interpretation that is in keeping with the evidence presented in this study. The pinjrapole, for example, illustrates that in certain segments of Indian society the ahiṃsā concept is an important determinant of human behavior, and one that is not dependent on any underlying cattle-protection function for its continued survival in modern India. The restrictions imposed by ahiṃsā on many areas of Jaina life, such as in diet, choice of occupation, or in personal habits, provide eloquent testimony to the influence that abstract or religious values can exert on personal, social, and economic behavior.

Indeed, the Jaina example illustrates vividly how an ideology can shape the nature of society, while at the same time being molded to that society's strategy for living. The result is a distinctive and unique way of life in which adherence to ahiṃsā is an integral part, and which makes nonsense of Harris's claim that the concept survives merely because of the material rewards it bestows on man and beast in India. Quite to the contrary, it is the central role that ahiṃsā plays in both Jaina philosophy and society that explains Jain support of the pinjrapole and Jain concern for the sanctity of animal life.

If this is so in Jainism, there is no reason to believe it should differ in Hinduism. Admittedly, ahiṃsā plays a slightly different role in Hindu thinking, yet historically it is the ahiṃsā concept applied specifically to cattle that has contributed to the emergence of the Hindu doctrine of the sanctity and inviolability of the cow, and sustained it through two millennia of Indian history. The ahiṃsā concept was a potent force in

Indian society centuries before the crystallization of the sacred cow concept, and its continued existence in India today can be accepted without resorting to functional materialistic explanations. Indeed, it is questionable whether ahiṃsā confers material rewards on either men or animals; it is dubious whether men are cognizant of these so-called rewards, surely a prerequisite for the concept deriving strength and power from them; and, finally, even if this were so, it is doubtful whether this interpretation of ahiṃsā can be expanded with equal validity to encompass the functioning of the sacred cow concept in modern times.

That behavior motivated by belief in concepts such as ahimsā and expressed in particular social traditions such as that of Jainism is not necessarily irrational or illogical, even though no material rewards may be forthcoming, is also apparent from the example of the pinjrapole. The sheltering and feeding of insects in the jīvat khān, for instance, appears at first sight to be a quaint and somewhat anachronistic custom in twentieth-century India. But what might appear to be odd behavior in a lay context becomes less so in terms of Jaina views of the universe, ideas of reincarnation and karma, and concept of soul. Again, Hindu attitudes toward the cow do not exactly parallel Jaina views of life; the sanctity of the cow is rationalized not only in terms of ahiṃsā, but also of a variety of factors ranging from mere sentiment to ideals of dharma. Yet it has been seen in this study that Hindu support of goshalas and pinjrapoles is, like Jain support, often justified on religious grounds. To dismiss religion and the logic inherent in religious systems as a determinant of individual and group behavior is to ignore an entire dimension of man and culture in India.

The historical development of animal homes in India, as far as it can be reconstructed, further emphasizes the religious nature of the institutions. All references to goshalas and pinjrapoles prior to the late nineteenth century would appear to relate to true pinjrapoles and temple goshalas, and even the majority of institutions established during the last century were founded primarily for cattle preservation rather than cattle development. The emergence of economic functions is of relatively late origin, dating back no more than, perhaps, a hundred years, while the dominance of economic activities in some goshalas and pinjrapoles is essentially a twentieth-century phenomenon.

It is no coincidence that this change in emphasis should occur at this particular time. For a variety of reasons, the rate of population growth in India during the late nineteenth and early twentieth centuries had been low. From the 1920s onward, however, we see a rapid burgeoning of the country's population, with annual growth rates exceeding 2.2 percent during the decade 1951–1961. This population expansion, with its resultant pressure on agricultural resources and increased demands on food supplies, was a contributory factor in the development of milk production and economic activities in animal homes. The transition from religious institution to economic also received further impetus from the newly formulated objectives of national economic planning in the post-Independence era and the policies implemented to accomplish these ends.

This recent focus on the utilitarian role of animal homes provides an interesting example of cultural evolution, religious institutions of considerable antiquity having developed economic functions in response to an increasingly unfavorable population-resource balance. It lends some credence to economic or ecological interpretations of animal husbandry practices in modern India, but also raises serious questions as to the validity of expanding these interpretations to include the entire sacred cow concept. If, for example, the economic activities of animal homes reflect the utilitarian nature of the sacred cow concept functioning under the techno-environmental conditions existing in India today, how can their presence be explained in the varying ecological contexts of past eras? Even one hundred years ago, the much smaller cattle population, more extensive areas of grazing and nineteenth-century agricultural technology made for a differing set of relationships within the Indian ecosystem. Yet then, as for centuries before, goshalas and pinjrapoles were present in India. They were, moreover, primarily religious institutions and still continue to function as such today.

In their historical development, therefore, the animal homes of India suggest that utilitarian interpretations of the sacred cow doctrine are less than adequate. Any attempt at developing a general theory of the sacred cow concept must not only explain the continued survival of their religious functions but also account for the existence of animal homes in the past. Economic or ecological interpretations do neither. Has there, in fact, been a fundamental change in the nature of the sacred cow concept so that it survives in India today merely because of the material benefits it supposedly bestows on men and animals? The evidence presented in this study indicates otherwise. In animal homes, religious considerations continue to be of importance, and where utilitarian functions have developed, these represent a recent change in emphasis in response to evolving environmental conditions and foreign cultural influences, not the total negation of the religious principles underlying their very existence.

Goshalas and pinjrapoles are, therefore, institutions deeply rooted in Indian history and culture. They incorporate beliefs and values originating in religious systems that are unique to India and that derive from views of the universe very different from those of the Judeo-Christian tradition. The origins, development, and functioning of these institutions make sense only in the historical, social, and philosophical context of Jainism and Hinduism, particularly with reference to the ahiṃsā concept and the doctrine of the sanctity and inviolability of the cow. Whether modern animal homes are survivals or imitations of the institutions of the past, they show a remarkable degree of persistence, having existed in some form for over two millennia. Over the centuries, they have evolved various forms and adapted to changing conditions, a process which is still apparent in India today. Perhaps at some time in the future they will have altered so radically that they may no longer be considered successors to the animal homes of the past. The institutions of today, however, are living reminders of the influences cultural values exert on human behavior. Given the techno-environmental conditions of contemporary India, it is unlikely that animal homes would exist without the frame of reference provided by the Hindu and Jain religious systems. But as long as the fundamental values of these religions survive in the subcontinent, institutions such as goshalas and pinjrapoles will remain a part of the Indian cultural scene.

APPENDIX A

LOCATIONAL REFERENCE MAPS

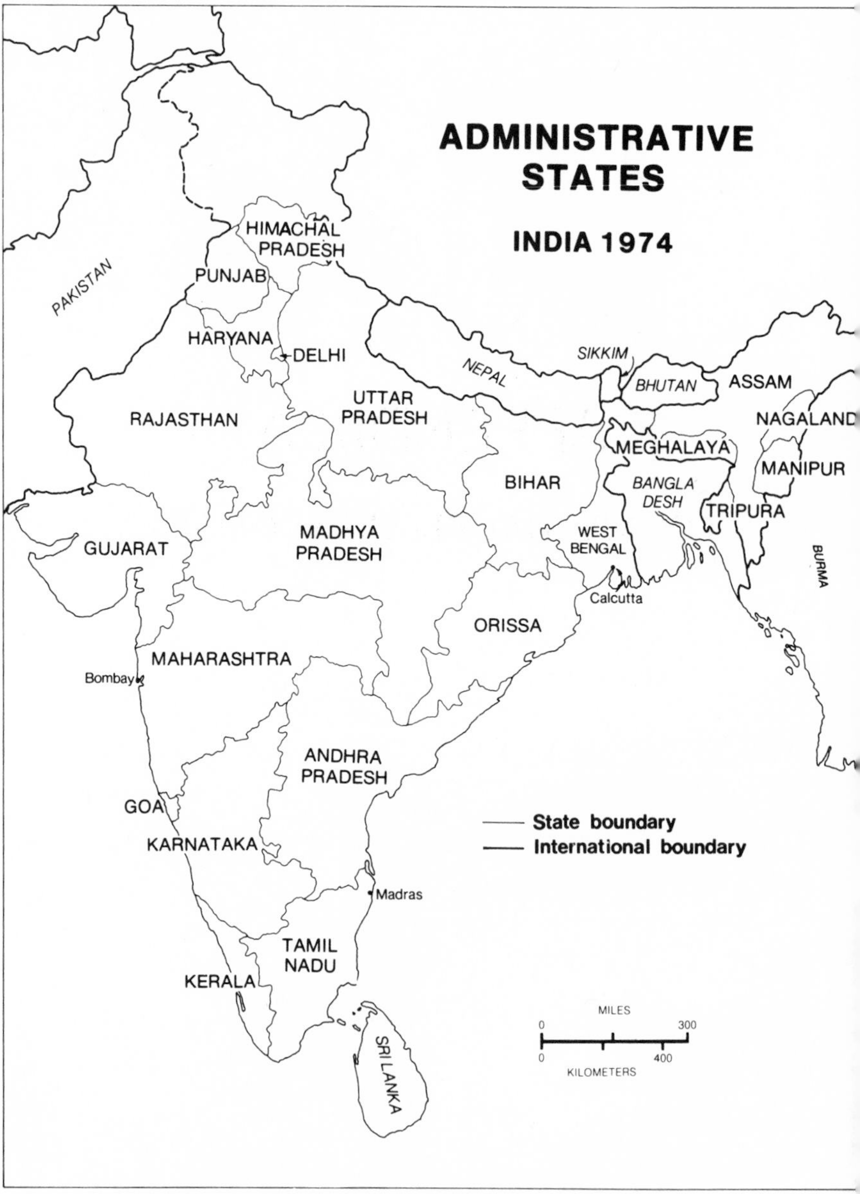
ADMINISTRATIVE STATES
INDIA 1974
PAKISTAN
HIMACHAL PRADESH
PUNJAB
HARYANA
DELHI
RAJASTHAN
UTTAR PRADESH
NEPAL
SIKKIM
BHUTAN
ASSAM
NAGALAND
MEGHALAYA
MANIPUR
BIHAR
BANGLA DESH
TRIPURA
WEST BENGAL
BURMA
Calcutta
GUJARAT
MADHYA PRADESH
ORISSA
MAHARASHTRA
Bombay
ANDHRA PRADESH
GOA
KARNATAKA
Madras
TAMIL NADU
KERALA
SRI LANKA
State boundary
International boundary
MILES
0
300
0
400
KILOMETERS

Map 13. *Administrative States of India, 1974*

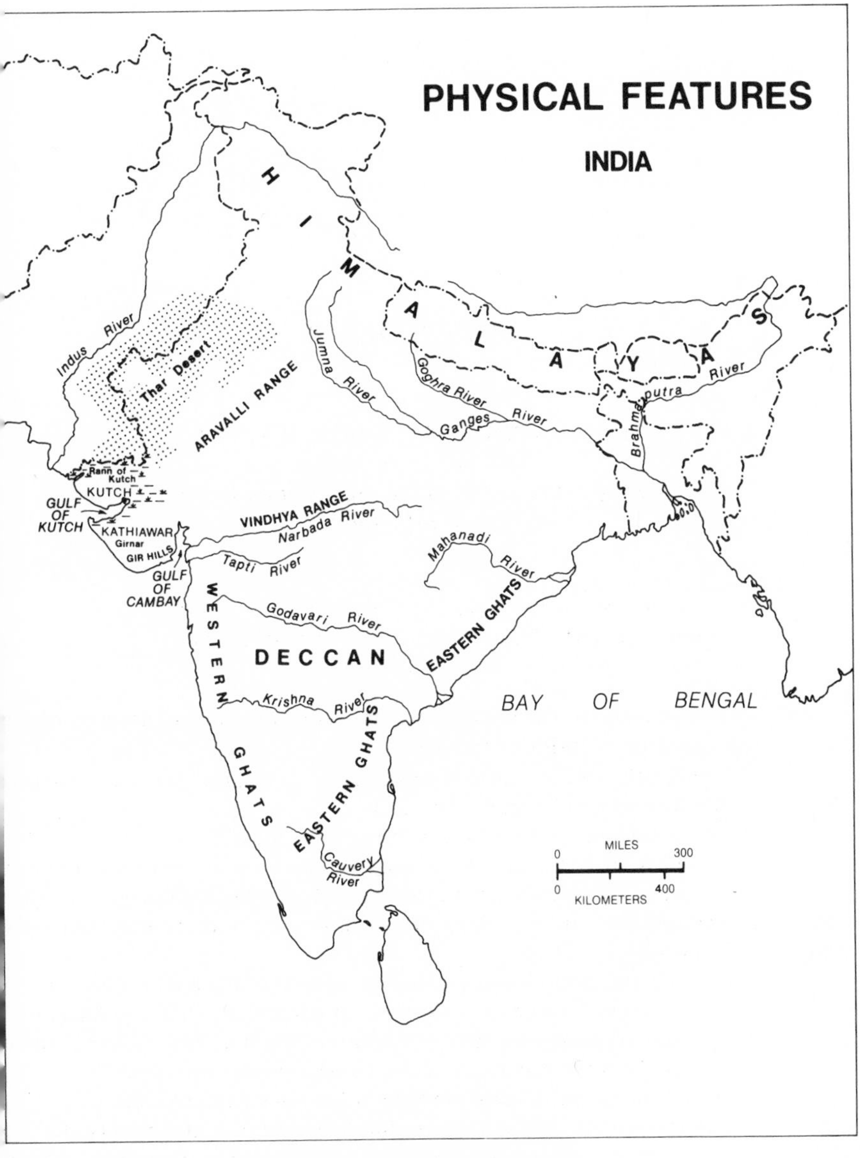
PHYSICAL FEATURES
INDIA
HIMALAYAS
Indus River
Thar Desert
ARAVALLI RANGE
Jumna River
Goghra River
Ganges River
Brahmaputra River
Rann of Kutch
KUTCH
GULF OF KUTCH
KATHIAWAR
Girnar
GIR HILLS
GULF OF CAMBAY
VINDHYA RANGE
Narbada River
Tapti River
Mahanadi River
Godavari River
DECCAN
WESTERN GHATS
EASTERN GHATS
Krishna River
Cauvery River
BAY OF BENGAL
MILES
0
300
0
400
KILOMETERS

Map 14. *Physical Features*

APPENDIX B

THE RAJASTHAN GAUSHALA ACT, 1960
(Act No. 24 of 1960)
(*Received the assent of the President on the 9th day of July, 1960*)

An
Act
to provide for better management, control and development of Gaushalas in the State of Rajasthan.

Be it enacted by the Rajasthan State Legislature in the Eleventh Year c the Republic of India as follows:—

1. *Short title, extent and commencement.*—(1) This Act may be calle the Rajasthan Gaushala Act, 1960.
   (2) It extends to the whole State of Rajasthan.
   (3) It shall come into force on such date as the State Governmen may, by notification in the Official Gazette, appoint.
2. *Definitions.*—In this Act, unless the subject or context otherwis requires,—
   (a) "Cattle" includes any domestic animal of the bovine species;
   (b) "Director" means the Director of Animal Husbandry, Rajasthan and includes any officer appointed by the State Government t perform the functions of the Director under this Act;
   (c) "Federation" means the Federation referred to in section 3;
   (d) "Gaushala" means a charitable institution established for th purpose of keeping, breeding, rearing and maintaining cattle o for the purpose of reception, protection and treatment of infirm aged or diseased cattle and includes a Pinjrapole or a Gosada where such cattle are kept;

(e) "Gaushala Development Officer" means the officer appointed as such by the State Government;

(f) "Registrar" means the Registrar of Gaushalas appointed under section 4, and includes an Assistant Registrar while performing the functions of the Registrar under this Act;

(g) "Regulation" means a regulation made by the Director under section 14;

(h) "Trustee" means a person or a body of persons, by whatever designation known, in whom the administration of a Gaushala is vested and includes any person who is liable as if he were a trustee.

3. *Establishment of the Federation.*—(1) As soon as may be possible, but not later than one year, after this Act comes into force, there shall be established in the State of Rajasthan a Federation to be called "The Rajasthan State Gaushala Federation."

(2) The Federation shall consist of the prescribed number of members elected by the trustees of the Gaushalas of the State in the prescribed manner at a meeting specially held for the purpose.

(3) Until the establishment of the Federation under this section, the Rajasthan Pinjrapole Gaushala Federation Jaipur which has been registered under the Societies Registration Act, 1860 (Central Act XXI of 1860), as adapted to Rajasthan, shall be the Federation for the purposes of this Act.

4. *Office of the Registrar and his staff.*—(1) The Gaushala Development Officer shall, by virtue of his office, be the Registrar of Gaushalas.

(2) Whenever necessary, the State Government may appoint an Assistant Registrar and such other staff on such salary, allowances and other conditions of service as it may determine.

5. *Trustees to apply for registration of and to furnish particulars relating to Gaushalas.*—(1) The trustee of every Gaushala shall, in the case of—

(a) a Gaushala established before the commencement of this Act, within three months of such commencement, and

(b) a Gaushala established after the commencement of this Act, within three months of such establishment, submit to the Registrar in the prescribed manner an application in the prescribed form for the registration of such Gaushala such particulars regarding the Gaushala as may be prescribed.

(2) The Registrar may, for reasons to be recorded in writing, extend the period of submitting such application and furnishing such statement.

(3) Upon receipt of an application under sub-section (1) the Registrar shall make such inquiry as he may deem necessary and may, if satisfied as to the correctness of the particulars furnished and the genuineness of the application, register the Gaushala in the register maintained under section 6 and issue a certificate of registration in the prescribed form.

6. *Register of Gaushalas.*—The Registrar shall maintain a register o Gaushalas in such form and containing such particulars as may b prescribed.
7. *Power of Registrar to hold inquiry.*—(1) The Registrar may at any tim either of his own motion or on the application of any person claiming t have an interest in the Gaushala or when required to do so by th Director or by the Federation, hold an inquiry to ascertain—
   (a) if the Gaushala is a Gaushala to which this Act applies;
   (b) the details of the property appertaining to such Gaushala;
   (c) the name and address of the trustee of such Gaushala;
   (d) the mode of succession to the office of the trustee of such Gaushal
   (e) the income and expenditure of such Gaushala; and
   (f) the sources of income of such Gaushala.
   (2) In every inquiry under this section the Registrar shall cause notic of such inquiry to be served on the trustee of the Gaushala an permit him to appear in person or through an agent duly authorise in writing.
   (3) On the conclusion of the inquiry the Registrar may pass such orde as he may deem fit as to the matters to which the inquiry relates.
8. *Trustee to furnish annual statement of changes in the particulars o Gaushala.*—(1) After the registration of a Gaushala under section 5 the trustee thereof shall submit to the Registrar a statement ever year in the month of June in the prescribed form and manner, showin changes in the particulars of the Gaushala which may have occurre during the preceding financial year.
   (2) The Registrar may, for reasons to be recorded, extend the period fo submission of the annual statement.
   (3) The Registrar may, after such inquiry as he may deem fit incorporate the changes in the register of Gaushalas.
9. *Maintenance of accounts and their audit.*—(1) The accounts of ever Gaushala which has been registered under section 5 shall be properl maintained and balanced each year. The accounts shall be audite annually by a person or persons approved by the State Government i this behalf. The auditor shall furnish copies of his audit note to th trustee of the Gaushala and to the Registrar within four months of th end of the accounting year or within such further time as the Registra may, for reasons to be recorded in writing, grant.
   (2) Every auditor acting under sub-section (1) shall have access to th accounts and to all books, vouchers and other documents an records in the possession or under the control of the trustee.
   (3) Within six months of the end of the year for which the accounts ar balanced, or within such further time as the Registrar may, fo reasons to be recorded in writing, grant, the trustee of ever Gaushala shall furnish to the Registrar a statement of the account in such form and containing such particulars as may be prescribed
10. *Inspection of Gaushalas.*—The Registrar or any person authorised b

him in this behalf may enter into and inspect any Gaushala or any place appertaining to such Gaushala for the purpose of satisfying himself that the provisions of this Act and the rules or the regulations made thereunder are duly complied with.

11. *Penalties.*—(1) If the trustee of a Gaushala fails or neglects to submit to the Registrar an application or a statement as required by section 5 or section 8 or submits an application or statement or furnishes particulars which he knows or has reason to believe to be false in any material particulars, such trustee shall, on conviction, be punishable with fine not exceeding one hundred rupees.

(2) If the trustee of a Gaushala fails or neglects to keep accounts or to furnish statement of accounts as required under section 9 or furnishes a statement which he knows or has reason to believe to be false in any material particulars, such trustee shall, on conviction be punishable with fine not exceeding five hundred rupees.

(3) If any person contravenes any other provision of this Act or any rule or any regulation made thereunder or fails to comply with any order made in pursuance of such provision, rule or regulation he shall, if no other penalty is provided elsewhere in this Act for such contravention, on conviction, be punishable with fine not exceeding fifty rupees.

(4) The court may, while passing an order of conviction and sentence under sub-section (1), (2) or (3), specify a period within which the person convicted shall comply with the provision of this Act or the rules or regulations made thereunder which may be found to have been contravened by him. If the person fails to comply with the order of the court within the specified period the court may also impose a fine not exceeding twenty rupees for every day of the period during which the default continues after the expiry of the period so specified:

Provided that, if such person satisfies the court that there was good and sufficient reason for his failure to comply with the order of the court within the period so specified, the court may, if it thinks fit, extend the period and may remit the whole or any part of the fine.

12. *Cognizance of offences.*—(1) No prosecution under this Act shall be instituted except on the complaint of the Registrar.

(2) No court inferior to that of a Magistrate of the Second Class shall try any offence under this Act.

13. *Power to make rules.*—(1) The State Government may make rules generally for the purpose of carrying into effect the provisions of this Act and in particular for prescribing all matters which may be, or are required to be, prescribed.

(2) All rules made under this Act shall be laid, as soon as may be after they are so made, before the House of the State Legislature, while it is in session, for a period of not less than fourteen days

which may be comprised in one session or in two successive sessions and, if, before the expiry of the session in which they are so laid or of the session immediately following, the House of the State Legislature makes any modification in any of such rules or resolves that any such rule should not be made, such rule shall thereafter have effect only in such modified form or be of no effect, as the case may be, so however that any such modification or annulment shall be without prejudice to the validity of anything previously done thereunder.

14. *Power to make regulations.*—(1) The Director may make regulations for the following matters, namely:—
   (a) the manner in which a Gaushala shall be managed, re-organised and developed;
   (b) skilled technical management of breeding work and segregation of such work from other activities of the Gaushala and the transfer of such work from urban to rural areas;
   (c) transport of breeding bulls from a Gaushala to any other place for breeding purposes;
   (d) veterinary treatment and inspection of cattle at Gaushala;
   (e) setting aside of cattle, both male and female, for breeding purposes.
   (2) Regulations made under this section shall be subject to the condition of previous publication and shall come into operation after they have been approved by the State Government.

15. *Bar of application of the Charitable and Religious Trust Acts, 1920.*—The provisions of the Charitable and Religious Trusts Act, 1920 (Central Act XIV of 1920) shall not apply to any Gaushala registered under this Act.

16. *Power to exempt.*—The State Government may, by notification in the Official Gazette, exempt any Gaushala or class of Gaushalas under special circumstances from all or any of the provisions of this Act.

APPENDIX C

Arrangement Entered Into for the Institution of the Bombay Panjrapole[1]

Under date the 1st Ashovud Sumvat 1890, corresponding with Saturday the 18th October 1834, the undersigned Shree Gosainjee Maharaj, the Shetias of the Hindoos, as also of the Parsees and rest of the Merchants of Bombay, have unanimously resolved upon to pay always Luwajums or fees on certain Mercantile Commodities on account of the undermentioned charitable purposes. The reason of this is as follows:

The Hon'ble East India Company has made a regulation to kill annually the dogs on the Island of Bombay, and has in conformity to that regulation carried on the practice of killing those animals and consequently quarrels frequently take place between the sepoys of the Hon'ble Company's Service and the Hindoo and Parsee (ryots); for the killing of animals in general is inconsistent with the religion of the Hindoos and the Parsees, and is a heinous act and requires atonement. Formerly petitions were several times submitted to Government praying for the prohibition of the practice, but to no purpose. This practice has therefore become a great nuisance. Having considered that these animals were undeservedly killed, and having resolved upon a remedy which secured their lives, and was in no way contrary to our religions, we further made a petition to the Hon'ble the Governor in Council in 1832, stating that if Government would prohibit the practice of killing dogs, we would cause them to be caught alive and sent to other ports and places. The Hon'ble

[1]A copy of this document was made available to the author by the management of the Bombay Pinjrapole.

the Governor in Council complied with this request, and intimated to us that if we continued to cause them to be caught alive and sent to other places, Government would not cause them to be killed, but that if we did not continue to do so, they would not discontinue the practice. For this reason the practice of catching dogs alive and sending them to other places has been adopted, and it has been maintained by spending great sums, but as there is no income for sustaining this expense, the practice in question cannot be continued. It is therefore necessary to make an arrangement for the expense of catching dogs alive and feeding them. Besides this, the Government peons catch stray cattle and give them great trouble: this is also repugnant to our religions, and as there are Panjrapoles in many other places, and as their expenses are maintained by import and export fees, we, the undersigned Hindoos and Parsees, have resolved upon establishing a similar Panjrapole in Bombay for the keep of stray cattle and other animals, by which many lives will be protected. This will be a very charitable act. The (Shawuks) promise to give for the proposed Panjrapole a place with suitable buildings, by raising a subscription among themselves. But as it is necessary to feed the animals, it is incumbent upon us to raise a fund for their maintenance. We, the Hindoo and Parsee Mahajuns, have therefore conjointly resolved upon paying fees on the undermentioned Mercantile Commodities, for this charitable purpose. The Hindoos, Parsees, etc. shall pay these fees. The particulars of the fees are as follows:

1st. On the Cotton which may be brought from the Continent to Bombay, the owner of it, and the person in whose name it may be registered at the Government Custom House, or the person who may act as an agent, shall pay one quarter of a rupee per Surat candy.

2nd. On the opium which may be brought to Bombay, the owner of it, and the person in whose name it may be registered at the Government Custom House, or the person who may act as an agent, shal pay one rupee per chest.

3rd. On the Bengal bags, China Chests, Siam (Soorkhas) o sugar, and other sorts of packages of sugar weighing one pecul or up wards, which may be brought to Bombay, the owner of it, and the persor in whose name it may be registered at the Government Custom House, o the person who may act as an agent shall pay one anna on every package

4th. On the tubs of sugar candy, Manilla sugar, and Mauri tius sugar, which may be brought to Bombay, the owner of it, and the person in whose name it may be registered at the Government Custom House, or the person who may act as an agent, shall pay half an anna pe package.

5th. On the Foreign Bills of Exchange, drawn either in Bombay or in the country, which may be discounted, the discounter shal pay a quarter anna per cent.

6th. Merchants who deal in pearls must pay quarter anna on hundred rupees' worth of pearls.

If any of the above articles be imported either by European or Musselman, and be registered in his name in the Government Custom House, the fees due on them according to the abovementioned rates shall be paid by their purchasers.

In this manner we have resolved upon paying fees on merchandise, at the rates above alluded to, for this charitable purpose. We have also resolved upon to commence the payment of fees from Monday the 2nd Kartick Sood Sumvut 1891, corresponding with 3rd November 1834. As long as dealings in the abovementioned commodities are carried on in the Island of Bombay, the Hindoo and Parsee dealers in those articles, Shroffs and other Merchants discounting Bills of Exchange, and the persons who may hereafter begin to trade, shall, after honestly consulting their accounts at home, pay annually the fees which may be due. It has been agreed that the expenses of this charitable purpose shall always be maintained. The fees in question shall be collected annually and entered with the donor's name and other particulars in the account books of the Panjrapole establishment. The expense shall thus be maintained. The following four persons are appointed to manage all the business:—

Wadiajee Shet Bomanjee Hormusjee

Sir Jamsetjee Jeejeebhoy

Shet Motechund Ameerchund, and

Shet Wakhatchund Khooshalchund.

In this manner the above four individuals have been appointed to manage this affair. They shall collect the fees from all places, and defray the expenses of the abovementioned charitable purpose. Any surplus, after defraying the expenses, shall be deposited on interest in a firm which may be fixed upon by the four managers.

The deposit shall be entered under the head of Panjrapole, as paid through the managers. The expenses of the establishment which may be required to keep correct accounts, to collect the fees, to clean the animals in the Panjrapole, as also expense of feeding the animals and of sending them to other ports and places when their number becomes great, and any other incidental expense, shall be defrayed from the income in question. The said four individuals shall conduct the business. There shall be six other persons subordinate to them, to conduct the business of the Panjrapole. The names are as follows:—

Shah Kullianjee Kanjee

Shah Khimchund Premchund

Shah Foolchund Kupoorchund

Shah Somjee Tarachund
Shah Dullall Gopaljee Kimjee, and
Shah Ramchund Govindjee

In this manner these persons have been appointed. When successo are appointed to them, persons who are honest, possess a good characte and shall pay attention to the business of the Panjrapole, shall b chosen. The authority of these persons shall be determined by the fo managers, and other respectable merchants dealing in the abovemе tioned commodities of the time being. The fees which have been fixe upon all mercantile commodities, Bills of Exchange, for the aboveme tioned charitable purpose, shall be annually collected, the remaind shall be credited under the head of Panjrapole. If the fund thus raise increases to excess, and if the four managers think it proper to appropr ate the surplus to some other charitable purposes, they may do so; co rect annual accounts of such appropriations shall be entered in th account books, which shall be signed annually by the managers. Th accounts shall also be published in newspapers for the information the public. The said managers shall conduct the business of the Panjr pole; but all the Hindoos and Parsees shall, as far as it lies in the power, render any assistance which may be required. The fees to th Panjrapole shall be annually paid as long as the Island of Bombay is existence, and the charitable Institution maintained. This settlement ha been unanimously made by all Hindoos, Parsees, Mahajuns, and other and shall be honestly observed. We, our Heirs and Executors, acknow edge these arrangements:

(Signed by the Merchants of Bombay)

APPENDIX D

Goshala and Pinjrapole Questionnaire

Date ________________________ State ________________________

District ____________________

A. *Location*

1. Name and address of the Institution.
2. Names and addresses of branches.
3. Where are the Institution's premises located? (e.g., city centers, suburbs, villages, etc.)

B. *Origins*

1. When was the Institution founded?
2. Who founded the Institution? Name

   Religion

   Caste/*jat*

   Occupation

   (If the Institution was founded by a committee, give the above data for each member.)
3. If any of the founders belonged to the Marwari, Gujarati or any other outside community, from what place did they or their families originate? When did they migrate?
4. What was the original purpose of the Institution?

C. *Animal Population*

1. *Cattle*

	Pedigree	Western Cross	Graded	Deshi (Country)	Total
a. Cows in milk					
b. Pregnant cows					
c. Dry cows					
d. Old, disabled or sick cows					
e. Breeding bulls					
f. Other bulls					
g. Working bullocks					
h. Useless bullocks					
i. Heifers (1–3 years)					
j. Males (1–3 years)					
k. Females below 1 year					
l. Males below 1 year					

Total ________

2. *Buffalo*

	Breed	Number
a. Cows in milk		
b. Pregnant cows		
c. Dry cows		

	Breed	Number
d. Old, disabled or sick cows		
e. Breeding bulls		
f. Other bulls		
g. Working bullocks		
h. Useless bullocks		
i. Heifers (1–3 years)		
j. Males (1–3 years)		
k. Females below 1 year		
l. Males below 1 year		

Total ___________

3. *Other domesticated animals* (goats, pigs, sheep, poultry, etc.)
4. *Other non-domesticated animals*
5. *Source of Animals*
   a. (i) Out of the present herd, how many animals were bought specifically to provide milk?
      (ii) In what year?
      (iii) What breeds were purchased?
      (iv) What was the cost?
      (v) From where did the funds come?
      (vi) From where were the animals purchased?
   b. (i) How many milch cows were donated to the Institution?
      (ii) By whom were they donated?
      (iii) Why were they donated?
   c. How many animals were bought to prevent them from being slaughtered?
   d. (i) How many useless animals were donated to the Institution?
      (ii) By whom were they donated?
   e. How many animals in the institution were given as "godān"?
   f. What is the admission fee charged for animals?
   g. Do arrangements exist for the Institution to take in cows during their dry period on payment of fees, but to return them to their owners when they are in milk?

h. Is there a "khuraki" system of boarding cows?
i. Are stray animals accepted by the Institution?
j. Are useless cows segregated from the milk herds?

D. Resources

1. a. How much land is owned by the Institution?
   b. Was it donated or purchased by the Institution?
   c. If donated, who donated it?
2. Where is the land located?
3. How much land is devoted to—
   a. grazing
   b. cultivation
   c. buildings and roads
4. Are the animals grazed along the roadside or in the jungle?
5. How much cultivated land is devoted to—
   a. fodder crops (specify crop type and acreage)
   b. cash crops (specify crop type and acreage)
6. How much of the agricultural production is sold on the open market?
7. How is the fodder used? (green, preserved as silage, hay, etc.)
8. a. What is the daily consumption of fodder in the Institution?
   b. How much fodder has to be purchased?
   c. What kind of fodder is purchased?
   d. At what cost?
9. a. How much land is irrigated?
   b. What crops are grown on the irrigated land?
   c. What is the method of irrigation?
10. a. What buildings are owned by the Institution?
    (i) offices
    (ii) cowsheds
    (iii) go-downs
    (iv) houses
    (v) shops
    b. Are any of these rented to the public?
11. What agricultural equipment is owned by the Institution?
12. How many animals can be accommodated in the Institution?

E. Production

1. Milk Production

a. What is the annual production of milk?
b. Where is it marketed?
c. At what price is it sold?
d. Is there a coupon system?
e. How many families purchase milk from the Institution?

2. Animal Production

a. What is the annual production of—
   (i) breeding bulls

(ii) bullocks
(iii) cows

b. Are animals bred specifically for sale?
c. What animals are sold?
d. At what price?
e. To whom are they sold?

3. *Breeding Policy*
a. Are western breeds being introduced into the Institution?
b. Is cross-breeding practiced?
c. Is artificial insemination used?
d. Are males castrated at the Institution?
e. Are the services of the Institution's breeding bulls available to the public? If so, at what fee?

F. *Veterinary Facilities*

1. Is there a veterinarian resident at the Institution?
2. If not, what veterinary facilities are available to the Institution?
3. Is immunization practiced by the Institution?
4. What diseases are prevalent among the animals in the Institution?
5. How are dead animals disposed of?
   a. buried
   b. sold
   c. removed by Chamars
   d. removed by the municipality
6. What is the cost entailed in the disposal of carcasses?

G. *Management*

1. Is management personal, or by committee?
2. If personal, what is the caste, community and occupation of the manager?
3. If by committee, what is the caste, community and occupation of the officers and managing committee of the Institution?
4. How many workers are employed by the Institution?
   a. Administrators
   b. Workers

H. *Finances*

1. *Income*
a. What is the total annual income of the Institution?
b. How much income is received from—
(i) sale of milk
(ii) sale of cattle
(iii) sale of hides and carcasses
(iv) servicing of cows

(v) sale of gobar
(vi) sale of agricultural products

c. How much income is received from—
(i) lag, britti or katauthi
(ii) dharmada (chanda)
(iii) other donations (please specify)

d. If a lag or similar cess is collected—
(i) on whom is it levied?
(ii) on what products is it levied?
(iii) at what rate is it levied?

e. How much income is received from—
(i) rent of buildings
(ii) rent of agricultural land
(iii) interest on loans and bank accounts

f. How much income is received as grants and subsidies from the Government?

g. Has the Institution participated in any Government Goshala Scheme in the past? If so, give details of any assistance provided under the scheme.

2. *Expenditure*

a. What is the total annual expenditure of the Institution?

b. What is the annual expenditure for—
(i) cattle feed
(ii) wages and salaries
(iii) maintenance
(iv) other

3. *Assets*

a. What is the total value of the assets of the Institution? (e.g., land, buildings, deposit accounts, cattle, etc.)

I. Religious Functions

1. Which of the following festivals are celebrated at the Institution?
a. Gopāṣṭamī
b. Goverdhan Pūjā
c. Janamaṣṭamī
d. Others (please explain)

2. Explain what happens during Gopāṣṭamī and any other festival held at the Institution.

3. Are there any temples or shrines on the Institution's premises? Dedicated to

Shiva
Vishnu
Krishna
Hanumān
Others

4. Have contributions ever been made to the Institution by people other than Hindus and Jains? (e.g., Buddhists, Moslems, Christians, etc.) If so, please explain when and why these contributions were made.

When this questionnaire is returned, please enclose a copy of the latest annual report of the Institution.

APPENDIX E

Goshala and Pinjrapole Survey

1974–1975

The following institutions were personally visited and surveyed by me during a period of field research in India from January 1974 to September 1975.

	Institution	*Location*	*District*	*Type*
ASSAM				
	1. Śri Gauhati Pinjrapole	Gauhati	Kamrup	vania
	2. Gerua Gram Seva Goshala	Gerua	Kamrup	private
	3. Nilachan Goshala	Kalipur	Kamrup	ashram
	4. Sarihatoli Dudhnut Bigrha Goshala	Nalbari	Kamrup	vania
	5. Śri Gopal Goshala	Dibrugarh	Lakhimpur	vania
	6. Śri Krishnan Govardhandhari Goshala	Chunpara	Sibsagar	vania
BIHAR				
	7. Gorakshan Sabha	Arrah	Bhojpur	vania
	8. Śri Adarsh Goshala	Buxar	Bhojpur	vania
	9. Śri Son Vaiganik Goshala	Dehri	Rohtas	vania
	10. Śri Krishna Goshala	Sasaram	Rohtas	vania

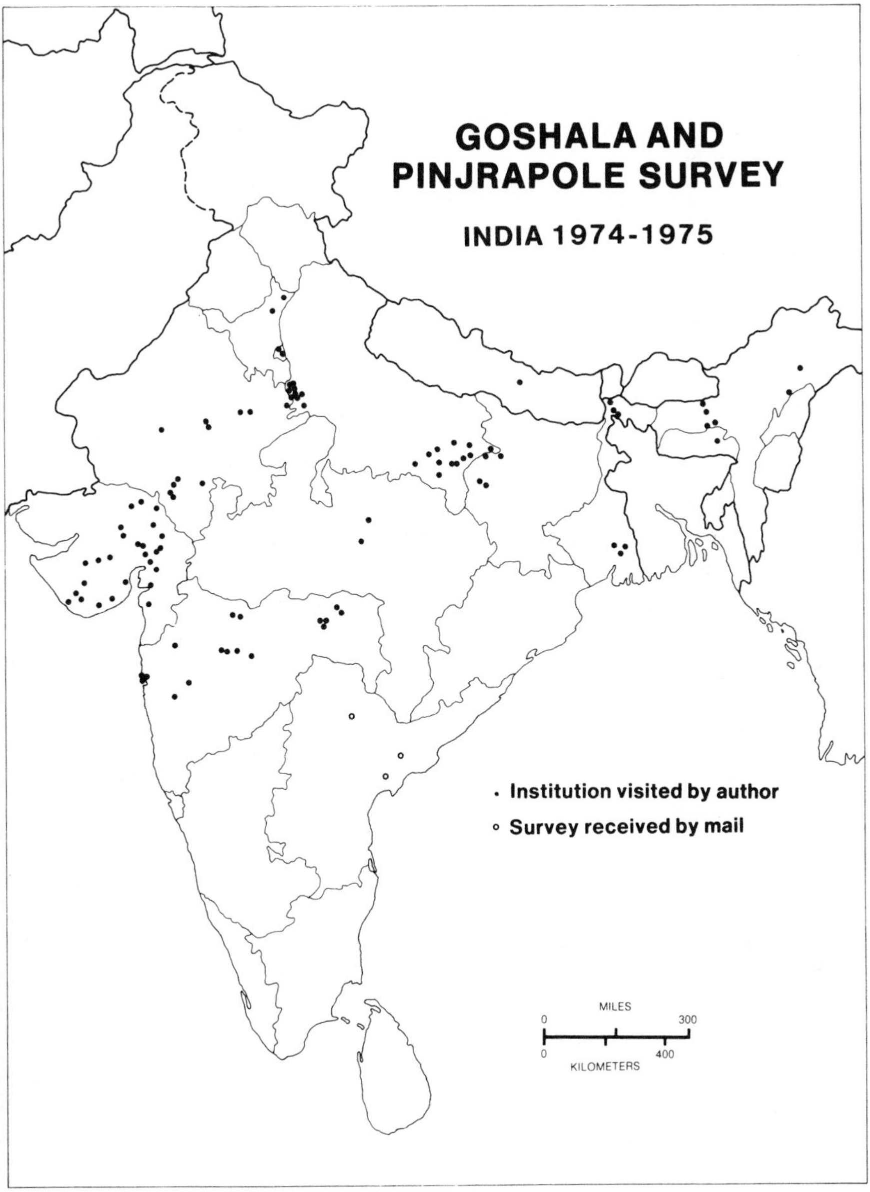

Map 15. *Goshala and Pinjrapole Survey*

	Institution	Location	District	Type
GUJARAT				
11.	Ahmedabad Pinjrapole Sanstha	Ahmedabad	Ahmedabad	pinjrapole
12.	Jagannath Mandir Goshala	Ahmedabad	Ahmedabad	temple
13.	Mandal Mahajan Pinjrapole	Mandal	Ahmedabad	pinjrapole
14.	Viramgram Mahajan Pinjrapole	Viramgram	Ahmedabad	pinjrapole
15.	Palanpur Pinjrapole	Palanpur	Banas Kantha	pinjrapole
16.	Śri Baroda Pinjrapole	Baroda	Baroda	pinjrapole
17.	Śri Bhavnagar Pinjrapole	Bhavnagar	Bhavnagar	pinjrapole
18.	Chhapriali Pinjrapole	Chhapriali	Bhavnagar	pinjrapole
19.	Broach Pinjrapole Goshala	Broach	Broach	pinjrapole
20.	Junagadh Pinjrapole Goshala	Junagadh	Junagadh	vania
21.	Virani Goshala	Shadagram	Junagadh	educational
22.	Tulsi Shyam Mandir Goshala	Tulsi Shyam Mandir	Junagadh	temple
23.	Visavadar Gorakshan Pinjrapole	Visavadar	Junagadh	vania
24.	Khoda Dhor Pinjrapole	Dakor	Kaira	pinjrapole
25.	Ranchorji Mandir Goshala	Dakor	Kaira	temple
26.	Khoda Dhor Pinjrapole	Kaira	Kaira	pinjrapole
27.	Mithabhai Gulabchand Pinjrapole	Kapadvanj	Kaira	pinjrapole
28.	Khoda Dhor Pinjrapole	Nadiad	Kaira	pinjrapole
29.	Śri Patan Mahajan Pinjrapole	Patan	Mehsana	pinjrapole
30.	Jetpur Mahajan Pinjrapole Goshala	Jetpur	Rajkot	pinjrapole
31.	Rajkot Mahajan Pinjrapole	Rajkot	Rajkot	pinjrapole
32.	Idar Pinjrapole Sanstha	Idar	Sabar Kantha	pinjrapole
33.	Prantij Pinjrapole	Prantij	Sabar Kantha	pinjrapole
34.	Surat Pinjrapole	Surat	Surat	pinjrapole

Institution	Location	District	Type
35. Chotila Pinjrapole	Chotila	Surendranagar	pinjrapole
36. Sayla Pinjrapole	Sayla	Surendranagar	pinjrapole
HARYANA			
37. Kaithal Goshala	Kaithal	Rohtak	vania
MADHYA PRADESH			
38. Pinjrapole Goshala	Jabbalpur	Jabbalpur	vania
39. Goshala Sanstha	Katni	Jabbalpur	vania
MAHARASHTRA			
40. Pinjrapole Gorakshan Sanstha	Ahmednagar	Ahmednagar	vania
41. Gorakshan Committee	Aurangabad	Aurangabad	vania
42. Mahavir Jain Goshala	Aurangabad	Aurangabad	vania
43. Mahavir Jain Goshala	Chikalthana	Aurangabad	vania
44. Śri Gorakshan Pinjrapole	Jalna	Aurangabad	vania
45. Bombay Gorakshan Mandali	Bombay	Bombay	vania
46. Bombay Pinjrapole Society	Bombay	Bombay	pinjrapole
47. Gorakshan League	Jogeshwari	Bombay	vania
48. Śri Ram Krishna Goshala	Bhusaval	Jalgaon	vania
49. Pinjrapole Sanstha	Jalgaon	Jalgaon	vania
50. Śri Gorakshan Sabha	Nagpur	Nagpur	vania
51. Śri Vyapari Gorakshan Sanstha	Virgaon	Nagpur	vania
52. Poona Pinjrapole Trust and Bhojpur Goshala	Poona	Poona	vania
53. Śri Nasik Panchayati Pinjrapole	Nasik	Nasik	vania
54. Sarve Seva Sangh Goshala	Gopuri	Wardha	Gandhian
55. Sevagram Goshala	Sevagram	Wardha	Gandhian
56. Gorakshan Mandal	Wardha	Wardha	vania
MEGHALAYA			
57. Śri Shillong Goshala	Shillong	Shillong	vania

	Institution	Location	District	Type
PUNJAB				
58.	Śri Krishna Goshala	Patiala	Patiala	vania
RAJASTHAN				
59.	Śri Pushkar Go-Adi Pasushala	Ajmer	Ajmer	vania
60.	Śri Budha Pushkar Goshala	Budha Pushkar	Ajmer	vania
61.	Śri Lalji Maharaj Mandir Goshala	Bharatpur	Bharatpur	temple
62.	Gurukul Goshala	Chitorgarh	Chitorgarh	educational
63.	Rajasthan Go-Seva Sangh	Jaipur	Jaipur	vania
64.	Pinjrapole Goshala	Sangameer	Jaipur	vania
65.	Siwandi Gate Goshala	Jodhpur	Jodhpur	vania
66.	Śri Eklingji Goshala	Śri Eklingji Mandir	Udaipur	temple
67.	Śri Nathji Goshala	Nathdwara	Udaipur	temple
68.	Go Seva Samiti	Udaipur	Udaipur	vania
69.	Goverdhan Vilas Goshala	Udaipur	Udaipur	temple
UNION TERRITORY				
70.	Delhi Pinjrapole Society	Delhi		vania
71.	Jain Pinjrapole	Delhi		pinjrapole
UTTAR PRADESH				
72.	Śri Goshala Society	Agra	Agra	vania
73.	Śri Pryag Gorakshani Samiti	Allahabad	Allahabad	vania
74.	Śri Krishna Goshala	Azamgarh	Azamgarh	vania
75.	Śri Gopal Goshala	Maunath Bhanjan	Azamgarh	vania
76.	Śri Lokman Tilak Goshala	Ballia	Ballia	vania
77.	Śri Gopal Krishna Goshala	Ghazipur	Ghazipur	vania

	Institution	Location	District	Type
78.	Śri Lakshmi Goshala	Saidpur	Ghazipur	vania
79.	Pinjrapole Pashupala Goshala	Jaunpur	Jaunpur	vania
80.	Śri Maqboo Hussein Goshala	Machlishsahr	Jaunpur	vania
81.	Śrimad Vallabh Goshala	Gokul	Mathura	vania
82.	Śri Krishna Goshala	Kosikalan	Mathura	vania
83.	Śri Krishna Goshala	Mathura	Mathura	vania
84.	Śri Nandgaon Goshala	Nandgaon	Mathura	vania
85.	Braj Mukat Mani Goshala	Udhaokund	Mathura	vania
86.	Hasanand Gochar Bhumi Trust Goshala	Vrindavan	Mathura	vania
87.	Śri Panchayati Goshala	Vrindavan	Mathura	vania
88.	Shyam Goshala	Vrindavan	Mathura	private
89.	Mirzapur Goverdhan Goshala	Mirzapur	Mirzapur	vania
90.	Śri Krishna Goshala	Bhadohi	Varanasi	private
91.	Banaras Hindu University Goshala	Varanasi	Varanasi	educational
92.	Śri Kashi Jiv-daya Bistarni Goshala and Pashushala	Varanasi	Varanasi	vania
WEST BENGAL				
93.	Calcattua Pinjrapole Society (CPS)	Calcutta	Calcutta	vania
94.	Darjeeling Goshala	Darjeeling	Darjeeling	vania
95.	CPS	Lilloah	Howrah	vania
96.	Śri Baikunthanath Pinjrapole Goshala	Jalpaiguri	Jalpaiguri	vania
97.	Śri Darjeeling-Siliguri Goshala	Siliguri	Siliguri	vania
98.	CPS	Sodepur	Twenty-four Parganas	vania
NEPAL				
99.	Śri Pashupati Goshala	Kathmandu		vania

	Institution	*Location*	*District*	*Type*

The following institutions returned completed questionnaires by mail:

ANDHRA PRADESH

	Institution	Location	District	Type
100.	Tenali Gosamrakshana Seva Sangham	Tenali	Guntur	vania
101.	Śri Goshala Committee	Nizamabad	Nizamabad	vania
102.	Eluru Gosamrakshana Samiti	Eluru	W. Godavari	vania

NOTES

I: Man, Culture, and Animals in India

1. Samuel Purchas, *Hakluytus Posthumus; or, Purchas His Pilgrimes, Contayning a History of the World in Sea Voyages and Lande Travells by Englishmen and Others*, 20 vols. (reprint of the 1625 ed.; New York: Macmillan, 1905–1907), 10:170.

2. Marco Polo, *The Book of Ser Marco Polo, the Venetian, Concerning the Kingdoms and Marvels of the East*, trans. and ed. Col Sir Henry Yule, 3d (Cordier) ed., 2 vols. (London: J. Murray, 1903), 2:341. Yule identifies Maabar as the twelfth-century Moslem name for the coastal areas of Tamil Nadu.

3. See, for example, H. J. Makhijani, *Gaushalas and Pinjrapoles* (New Delhi: Central Council of Gosamvardhana, 1963); Harbans Singh, *Gaushalas and Pinjrapoles in India* (New Delhi: Central Council of Gosamvardhana, 1955); Sardar Datar Singh, *Reorganisation of Gaushalas and Pinjrapoles in India* (New Delhi: Ministry of Agriculture, 1948).

4. For a recent survey of studies concerning man and animals, see Frederick J. Simoons, "Contemporary Research Themes in the Cultural Geography of Domesticated Animals," *Geographical Review* 64 (1974):557–576.

5. Frederick J. Simoons, *A Ceremonial Ox of India* (Madison, Wisconsin: University of Wisconsin Press, 1968).

6. Frederick J. Simoons, *Eat Not This Flesh* (Madison, Wisconsin: University of Wisconsin Press, 1961), p. 9.

7. Frederick J. Simoons, "The Traditional Limits of Milking and Milk Use in Southern Asia," *Anthropos* 65 (1970):547–593. See also, "The Determinants of Dairying and Milk Use in the Old World: Ecological, Physiological and Cultural," *Ecology of Food and Nutrition* 2 (1973):83–90.

8. Frederick J. Simoons, "Fish as Forbidden Food: The Case of India," *Ecology of Food and Nutrition* 3 (1974):185–201.

9. Robert Hoffpauir, "India's Other Bovine: A Cultural Geography of the Water Buffalo" (Ph.D. dissertation, University of Wisconsin, 1974).

10. Richard P. Palmieri, "Domestication and Exploitation of Livestock in the Nepal Himalaya and Tibet: An Ecological, Functional and Culture Historical Study of Yak and Yak Hybrids in Society, Economy, and Culture" (Ph.D. dissertation, University of California, Davis, 1976).

11. Molly Debysingh, "Poultry and Cultural Distributions in India" (Ph.D. dissertation, Syracuse University, 1970).

12. Gary S. Dunbar, "Ahimsa and Shikar: Conflicting Attitudes towards Wildlife in India," *Landscape* 19, no. 1 (1970):24–27.

13. India (Republic), Ministry of Food, Agriculture, Community Development and Cooperation, Directorate of Economics and Statistics, *Indian Livestock Census, 1961* (Delhi: Manager of Publications, 1969).

14. India (Republic), Ministry of Planning, Department of Statistics, Central Statistical Organization, *Statistical Abstract. India, 1972* (Delhi: Controller of Publications, 1974).

15. India (Republic), *Domestic Fuel Consumption in Rural India* (New Delhi: National Council of Applied Economic Research, 1965), p. 28.

16. The inclusion of a ban on cow slaughter was the subject of heated debate at the time of the drafting of the Constitution of India during the late 1940s. Hindu traditionalists pressed for a complete ban on the slaughter of all common cattle and sought to incorporate it into the section of the Constitution dealing with fundamental rights, which would have made it enforceable in the country's courts of law. However, secularists and religious minorities opposing this move succeeded in having the ban included in the Directive Principles of State Policy section of the Constitution instead. The Directives set out basic principles to be followed by both central and state governments in the making of laws, but they themselves were not legally enforceable. The ban on cow slaughter is thus written into the Indian Constitution as Article 48:

> Organisation of agriculture and animal husbandry.—The State shall endeavour to organise agriculture and animal husbandry on modern and scientific lines and shall, in particular, take steps for preserving and improving the breeds, and prohibiting the slaughter, of cows and calves and other milch and draught animals.

Most of the states of India have followed this directive in enacting, or promising to enact, total bans on the slaughter of all cattle. In West Bengal, however, the ban is only partial, and Kerala, with its large Christian and Moslem minorities, has no restrictions on the slaughter of cattle. It was largely against these two states that Vinoba Bhave's fast was directed. The vague wording of Article 48 and the varying interpretations of antislaughter legislation enacted at both the state and local levels has led to litigation that has been taken all the way to the Supreme Court of India (Frederick J. Simoons, "The Sacred Cow and the Constitution of India," *Ecology of Food and Nutrition* 2 [1973]:281–295).

17. It is by no means coincidental that West Bengal and Kerala, the only two Indian states that do not have complete bans on the slaughter of cattle, are regional strongholds of the Communist Parties of India. Both states have in the recent past been governed by coalitions led by the Communist Party of India (Marxist).

18. W. Norman Brown, "The Sanctity of the Cow in Hinduism," *The Madras University Journal* 28 (1957):29–49; reprinted in *Economic Weekly* (Bombay) *16* (1964):245–255.

19. See, for example, William Crooke, "The Veneration of the Cow in India," *Folk-Lore* 23 (1912):275–306; Hermann Jacobi, "Cow (Hindu)," pp. 224–226 in James Hastings, ed., *Encyclopaedia of Religion and Ethics* 4 (1955); and Arthur A. Macdonell, *Vedic Mythology* (Varanasi: Rameshwar Singh, 1896).

20. B. R. Ambedkar, *The Untouchables* (New Delhi: Amrit Book Company, 1948), pp. 116–121.

21. Paul Diener, Donald Nonini and Eugene E. Robkin, "The Dialectics of the Sacred Cow: Ecological Adaptation vs. Political Appropriation in the Origins of India's Cattle Complex," *Dialectical Anthropology* 3 (1978):221–241.

22. Marvin Harris, "The Myth of the Sacred Cow," pp. 217–228 in Anthony Leeds and Andrew P. Vayda, eds., *Man, Culture and Animals* (Washington, D.C.: American Association for the Advancement of Science, 1965) and "The Cultural Ecology of India's Sacred Cattle," *Current Anthropology* 7 (1966):51–66.

23. Julian H. Steward, *Theory of Culture Change: The Methodology of Multilinear Evolution* (Urbana, Illinois: University of Illinois Press, 1955).

24. Harris, op. cit., 1965:226.

25. Paul Diener and Eugene E. Robkin, "Ecology, Evolution and the Search for Cultural Origins: The Question of Islamic Pig Prohibition," *Current Anthropology* 19 (1978):493–540.

26. Diener, Nonini and Robkin, op. cit., 1978; Alan Heston, "An Approach to the Sacred Cow of India," *Current Anthropology* 12 (1971):191–209; V. M. Dandekar, "India's Sacred Cattle and Cultural Ecology," *Economic and Political Weekly* 4 (1969):1559–1567; John W. Bennett, "On the Cultural Ecology of Indian Cattle," *Current Anthropology* 8 (1967):251–252.

27. Marvin Harris, *Cannibals and Kings* (New York: Random House, 1977), p. 147.

28. Frederick J. Simoons, "Questions in the Sacred-Cow Controversy," *Current Anthropology* 20 (1979):467–493.

29. Marvin Harris, op. cit., 1966:52.

30. John W. Bennett, "Comment on: An Approach to the Sacred Cow of India, by Alan Heston," *Current Anthropology* 12 (1971):197.

31. Simoons, op. cit., 1979:468.

32. Max Weber, *The Protestant Ethic and the Spirit of Capitalism* (New York: Charles Scribner's Sons, 1958), p. 25.

33. Max Weber, *The Religion of India,* trans. and ed. H. H. Gerth and D. Martindale (Glencoe, Illinois: The Free Press, 1958), p. 111.

34. J. A. Dubois, *Hindu Manners, Customs and Ceremonies*, trans. and ed. Henry K. Beauchamp, 3d ed. (Oxford: The Clarendon Press, 1906), p. 96.

35. Vera Anstey, *The Economic Development of India* (London: Longmans, Green and Co., 1936), p. 47.

36. Vikas Mishra, *Hinduism and Economic Growth* (London: Oxford University Press, 1962), p. 206.

37. K. William Kapp, *Hindu Culture, Economic Development and Economic Planning* (Bombay: Asia Publishing House, 1963), p. 64.

38. Kusum Nair, *Blossoms in the Dust* (New York: Frederick A. Praeger, 1962).

39. Milton Singer, "Cultural Values in India's Economic Development," *An-*

nals of the American Academy of Political and Social Science* 305 (1956):81–91.

40. M. D. Morris, "Values as an Obstacle to Economic Growth in South Asia: An Historical Survey," *Journal of Economic History* 27 (1967):588–607.

41. Calvin W. Schwabe, "The Holy Cow—Provider or Parasite? A Problem for Humanists," *Southern Humanities Review* 13, no. 3 (1978):251–278.

42. Marshall Sahlins, *Culture and Practical Reason* (Chicago: University of Chicago Press, 1976), pp. 170–204.

43. Ibid., p. 211.

44. For a further discussion of this question, see, for example, Sahlins, op. cit., 1976; John W. Bennett, *The Ecological Transition: Cultural Anthropology and Human Adaptation* (New York: Pergamon Press, 1976); Richard Newbold Adams, *Energy and Structure: A Theory of Social Power* (Austin: University of Texas Press, 1975).

II: Goshalas and Pinjrapoles—Forms in Modern India

1. Harbans Singh, *Gaushalas and Pinjrapoles in India* (New Delhi: Central Council of Gosamvardhana, 1955), p. 7.

2. P. A. Xavier, "The Role of Pinjrapoles in the Animal Welfare Work of India," *Animal Citizen (Madras)* 4, no. 3 (April–May–June 1967):35. Yule and Burnell, on the other hand, provide an alternate interpretation in *Hobson-Jobson: A Glossary of Colloquial Anglo-Indian Words and Phrases, and of Kindred Terms, Etymological, Historical, Geographical and Discursive* (London: Routledge and Kegan Paul, 1903). They suggest that the word's origins lie in the Gujarati terms *pinjra* (cage) and *pola* (the sacred bull released in the name of Shiva).

3. Satish C. Das Gupta, *The Cow in India*, 2 vols. (Calcutta: Khadi Pratisthan, 1945), 1:565. This cooperative community effort may well be a forerunner of the practice common in northern India today of leaving the village cattle in the care of a herder, usually an Āhīr or Goāla, with cash or grain payments being made by the owners of the cattle.

4. Sardar Datar Singh, *Reorganisation of Gaushalas and Pinjrapoles in India* (New Delhi: Ministry of Agriculture, 1948), p. 7.

5. K. K. Shukla, "Goshala Development Scheme. Gujrat State," *Gosamvardhana* 11, no. 1 (April 1963):31.

6. Y. M. Parnerkar, "New Vistas of Activity for Goshalas. Certified Milk Production," Editorial, *Gosamvardhana* 11, no. 11 (February 1964):2.

7. M. K. Gandhi, *India's Food Problem* (Ahmedabad: Navajivan Publishing House, 1960), p. 31.

8. M. K. Gandhi, *How to Serve the Cow* (Ahmedabad: Navajivan Publishing House, 1954), p. 31.

9. H. J. Makhijani, *Gaushalas and Pinjrapoles* (New Delhi: Central Council of Gosamvardhana, 1963).

10. The nilgai, the so-called blue bull, is actually a species of antelope (*Boselaphus tragocamelus*) but is often treated by Hindus with some of the respect normally reserved for the cow. A farmer in Uttar Pradesh related to this author how the local Moslem population, who would hunt and eat the animal, called it *nilgora* (literally "blue horse") rather than nilgai ("blue cow") to avoid offending the sensibilities of the Hindus. Attitudes toward the animal are, however, ambivalent. I am personally aware of instances in Rajasthan in which Hindu villagers have approached local *śikaris* (hunters) to shoot nilgai that have been feeding in their fields.

11. Jainism emerged during the sixth century B.C. as a reform movement opposing the excesses of Brahmanical Hinduism. Many scholars, and the Jains themselves, would ascribe an even greater antiquity to the religion but it was Mahāvīra, the twenty-fourth tīrthaṅkara and a contemporary of Buddha, who first gave the religion prominence in India. As is true with most religions, Jainism experienced the emergence of many sects and subsects, but the most important schism was that which occurred in the first century A.D. resulting in the division of Jains into Swetambara and Digambara. This, however, was based on a point of monastic discipline and was never reflected in any major doctrinal differences. Jainism never spread beyond India and today is followed by only 0.5 percent of the Indian population.

12. Samuel Purchas, *Hakluytus Posthumus; or, Purchas His Pilgrimes, Contayning a History of the World in Sea Voyages and Lande Travells by Englishmen and Others*, 20 vols. (reprint of the 1625 ed.; New York: Macmillan, 1905–1907), 10:170,182.

13. Alexander Burnes, "Art. VIII. Notice of a remarkable Hospitable for Animals at Surat. By Lieut. Alexander Burnes of the Bombay Military Establishment: being an Extract from a Manuscript Journal," *Journal of the Royal Anthropological Society of Great Britain and Northern Ireland* 1 (1834):96.

14. Literally "seeing," the word *darśan* is used to describe the entry of the worshipper into the presence of the deity, in this instance of the cow. The teachings of Vallabhāchārya emphasize the manifestation of God's grace through the sight of his image, and hence *darśan* is used among Krishna followers in preference to *Pūjā*, the general term for worship among Hindus.

15. A ceremony performed during worship by waving a ghī lamp in a circular motion before the head of the image of the deity.

16. M. K. Gandhi, *How to Serve the Cow* (Ahmedabad: Navajivan Publishing House, 1954), pp. 3–4.

17. The goshala at Sabarmati is now defunct, that at Uruli Kanchan is a modern dairy and agricultural research station, but the goshala at Sevagram is still in existence today.

III. Distributions, Locations, and Spatial Hierarchies

1. A simple correlation between numbers of pinjrapoles in a district and percentage of the population belonging to the Jain religion yields a coefficient (r) of +0.689.

2. Marwari refers to a dialect spoken in the former princely state of Marwar in what is now Rajasthan, but is used loosely throughout India to refer to Rajasthani businessmen who have migrated to all corners of the land in pursuit of their business activities.

3. According to local tradition, before the Ranchorji temple was built an old man from Dakor named Barani used to make the journey to Dwarka for darśan of Krishna. When he became too old to make the pilgrimage on foot, he prayed to Ranchorji (Krishna) to visit him in his home. Krishna told him that he should go to Dwarka with an ox cart and that he would return with him to Dakor. The old man did as he was told but was unable to drive back, so went to sleep in his cart. When he awoke, he was back in Dakor with the Krishna mūrti, having been driven back by the god himself. Some Goghaly Brahmans, the *pūjāris* (priests) at Dwarka, came to Dakor and accused Barani of the theft of the statue of Krishna from Dwarka, but decided that if he gave Ranchorji's weight in gold, the statue could remain. Barani was a poor man and prayed to Krishna for aid, to which the god replied that his weight would be that of Barani's wife's gold nose ornament. The statue was placed on some scales and to everyone's astonishment was balanced by the nose ornament of Barani's wife. The image of Ranchorji remained

in Dakor and the temple was built to house it. Today, at the temple there is a set of silver scales (*tūla*) six feet high which supposedly dates to the time of the founding of the temple. It is used for *tūladān*, the ceremony of weighing babies, children and even adults against wheat, sugar, and other goods that are then donated to the temple goshala.

4. In a land abounding with gods, earth spirits, snake spirits, tree spirits, and numerous other minor deities, the major gods of the Hindu pantheon have their own temples dedicated to them. In Gujarat, as indeed in the rest of India, Vishnu and Shiva are the most important of all the gods. The once all-powerful Brahmā, the Creator, has been totally eclipsed by them and today, in all of India, there is only one temple, at Pushkar in Rajasthan, dedicated to him. Lesser gods also receive homage. Thus, in Gujarat, temple goshalas are also associated with the worship of Hanumān, the monkey-god who was the helper of Rāma in the Rāmāyaṇa epic (Bala Hanumanji Mandir, Vana, District Surendranager); with the ancient cult of the Mother Goddess, worshipped as the source of *śakti*, the potency of her male counterpart (Holmataji Mandir, Wankaneer, District Rajkot); and with the sun-god Sūrya or Sūraj (Nava Suraj Deva Mandir, Devasir, District Surendranager).

5. Prior to 1947, goshalas existed in what is now Pakistan and Bangla Desh but few, if any, survived the upheavals of Partition. One refugee from West Pakistan who used to manage a goshala claimed that many institutions tried to drive their animals into Indian territory at the time of Partition, though few apparently succeeded in doing so. To the best of my knowledge, today no institutions are to be found outside India and Nepal.

6. For a further discussion of Krishna, see pp. 65–67 of text.

IV: Goshalas and Pinjrapoles—Origins and Historical Development

1. Hahn's ideas concerning the domestication of cattle and of other animals are found in *Die Haustiere und ihre Beziehungen zur Wirtschaft des Menschen* (Leipzig: Duncker & Humblot, 1896).

2. Charles A. Reed, "The Pattern of Animal Domestication in the Prehistoric Near East," in Ucko and Dimbleby, eds., *The Domestication and Exploitation of Plants and Animals* (Chicago: Aldine Publishing Co., 1969), p. 372.

3. James Mellaart, *Çatal Hüyük* (London: Thames and Hudson, 1967). Mellaart dates a shrine involving a bull of a wild species found at Level X to 6385 B.C.±101.

:ult activities at Çatal Hüyük were not restricted to bovines alone, for remains and epresentations of wild boar, rams, vultures, and other animals are found in ituations indicating possible religious associations. One feature of cult activity at ne site, however, is the prominence of the bull, possibly as a masculinity symbol in fertility cult.

4. Dyaus is the great father god of the Vedas, Rudra a malevolent storm-god nd Indra a god of war who is also a benign weather-god.

5. Elise J. Baumgartel, *The Cultures of Prehistoric Egypt* (London: Oxford niversity Press, 1960–65), 2:144.

6. E. O. James, *The Ancient Gods* (New York: G. P. Putnam's Sons, 1960), p. 82.

7. Lewis Bayles Paton, "Ishtar," in Hastings, ed., *Encyclopaedia of Religion and thics* (New York: Charles Scribner's Sons, 1955), 2:430. Ishtar is also called creatress of the creatures" and "creatress of all things" which Paton suggests ccounts for the wild animals roaming freely in the park of Atargatis at Hierapolis escribed by Lucian (*De Dea Syria*, 28f., 39ff.). One wonders whether the hunting arks and animal enclosures of the ancient world and perhaps even the modern zoo night not ultimately be traced back to such beginnings. Some aspects of Ishtar eem to have been assumed in India by Shiva, who in addition to his incarnation s Pashupatinath, "Lord of the Beasts," is also, like Ishtar, the destroyer of life.

8. H. R. Hall and C. L. Woolley, *Ur Excavations*, Vol. I, *Al-Ubaid* (London:)xford University Press, 1927), p. 143.

9. Sir Charles Leonard Woolley, *Ur: The First Phases* (New York: Penguin ooks, 1946), p. 28.

10. W. Norman Brown, "The Sanctity of the Cow in Hinduism," *The Madras Jniversity Journal* 28 (1957):29.

11. See, for example, C. J. Gadd, "Seals of Ancient Indian Style found at Ur," *roceedings of the British Academy* 17 (1932):191–210; F. A. Durrani, "Stone /ases as Evidence of Connection between Mesopotamia and the Indus Valley," *Ancient Pakistan* 1 (1964):51–96; E. C. L. During-Caspers, "Further Evidence for :ultural Relations between India, Baluchistan, Iran and Mesopotamia in Early)ynastic Times," *Journal of Near Eastern Studies* 24 (1965):53–56; B. Buchanan, A Dated Seal Impression Connecting Babylonia and Ancient India," *Archaeology* 20 (1967):104–107; C. C. Lamberg-Karlovsky, "Trade Mechanisms in Indus-Mesopotamian Interrelations," *Journal of the American Oriental Society* 92, no. 2 1972):222–229 and George F. Dales, "Shifting Trade Patterns between the ranian Plateau and the Indus Valley in the Third Millennium B.C.," pp. 67–78 in *Colloques Internationaux du Centre National de la Recherche Scientifique. No.*

567. *Le Plateau Iranien et L'Asie Centrale des Origines à la Conquête Islamiqu* 1976.

12. Sir John Marshall, *Mohenjo-Daro and the Indus Civilisation* (Londo Arthur Probsthain, 1931), 2:649.

13. Ernest Mackay, *Further Excavations at Mohenjo-Daro* (Delhi: Manager Publications, 1937–38) and *Chanhu-Daro Excavations 1935–1936*, America Oriental Series, vol. 20 (New Haven, Connecticut: American Oriental Societ 1943).

14. Om Prakash, *Food and Drink in Ancient India* (Delhi: Munshi Ra Manohar Lal, 1961), p. 4.

15. The interpretation of this symbol which occurs repeatedly on the seals the Indus Valley is a matter of conjecture. Some authors view it as a feedin trough and therefore indicative of domestication or of animals kept in captivit Variations in the form of the symbol occur, and it has been interpreted by othe as a cult object. Simoons notes that it has been linked to offering stands used i certain fertility rites in central India today and also to offering platforms to spirit among the hill peoples of Assam. (Frederick J. Simoons, *A Ceremonial Ox of Indi* [Madison, Wisconsin: University of Wisconsin Press, 1968], p. 256).

16. A. L. Basham, *The Wonder That Was India* (London: Sidgewick an Jackson, 1954), p. 23.

17. E. O. James, *Myth and Ritual in the Ancient Near East* (New York: Frederic A. Praeger, 1958), p. 134.

18. Our understanding of the Harappan religion and of the subsequent histori cal development of Hinduism has been greatly influenced by Marshall, bu several authors question his identification of the figure shown on the seal i figure 7c as a Proto-Shiva. Marshall based this view on (1) the figure being three faced, as Shiva himself is represented in later Indian art; (2) the yoga-like postur of the figure depicted on the seal conforming to Shiva's reputation as *Mahāyog* the great yogi and king of ascetics; (3) the array of animals around the figure o the Indus Valley seal recalling Shiva's role as Pashupati, Lord of the Beasts; an (4) the horned headdress on the Proto-Shiva figure being indicative of divinity an also being the source of Shiva's emblem, the *triśūl* or trident. Doris Srinivasa argues, however, that the archaeological and later textual evidence does no support Marshall's interpretation, and that the figure represents a divine bull man, possibly a deity of fertility and abundance, but certainly not an early form c Shiva ("The So-called Proto-Śiva Seal from Mohenjo-Daro: An Iconologica Assessment," *Archives of Asian Art* 29 [1975–76]:47–58). See also H. P. Sullivan "A Re-examination of the Religion of the Indus Civilization," *History of Religion* 4(1964):115–125, and Alf Hiltebeitel, "The Indus Valley 'Proto-Śiva,' Re-examine

rough Reflections on the Goddess, the Buffalo, and the Symbol of *vāhanas*," *nthropos* 73 (1978):767–797.

19. Asko Parpola, S. Koskenniemi, S. Parpola and P. Aalto, *Further Progress in e Indus Script Decipherment* (Copenhagen: The Scandinavian Institute of sian Studies, Special Publications no. 3, 1970), p. 9.

20. James, op. cit., (1958):135.

21. Brown, op. cit., p. 30.

22. *Sacred Books of the East* (*SBE*) 44:191, 26.8.

23. For a discussion of this topic, see S. A. Dange, *Pastoral Symbolism from e Ṛgveda* (Poona: Poona University Press, 1970).

24. *SBE* 42:105, 373.

25. *Ṛg Veda* 1.32.2.

26. Dange, op. cit., p. 45.

27. For recent studies on Shaivism, see J. Gonda, *Viṣṇuism and Śivaism: A omparison* (London: Athlone Press, 1970); Mariasusai Dhavamony, *Love of God ccording to Śaiva Siddhanta: A Study in the Mysticism and Theology of Śaivism*)xford: Clarendon Press, 1971); David N. Lorenzen, *The Kāpālikas and Kālā-ukhas: Two Lost Śaivite Sects* (Berkeley and Los Angeles: University of Califor-ia Press, 1972); and Wendy Doniger O'Flaherty, *Asceticism and Eroticism in the Iythology of Śiva* (London: Oxford University Press, 1973).

28. Brown, op. cit., p. 42.

29. A. A. Macdonell and A. B. Keith, *Vedic Index of Names and Subjects* London: John Murray, 1912), 2:146; and A. A. Macdonell, *Vedic Mythology* Varanasi: Rameshwar Singh, 1896), p. 151.

30. Op. cit., p. 33.

31. A. B. Keith, *Religion and Philosophy of the Vedas* (Cambridge: Harvard niversity Press, 1925), p. 192.

32. Op. cit., p. 35.

33. The head of one of the four *mats*, or monasteries, established at the ardinal points of India (Badrinath, Shringeri, Dwarka, and Puri) by the philos-pher, Śaṅkara, at the beginning of the ninth century, and a Hindu religious

leader of some considerable standing. The Śaṅkarāchārya is also a major force i the current cow-protection movement in India which seeks a total ban on th slaughter of cattle in the country.

34. Macdonell and Keith, op. cit., 1:232, 240.

35. Op. cit., p. 102ff.

36. *SBE* 14:117.

37. *SBE* 14:249.

38. *SBE* 2:276, 14:311.

39. *SBE* 25:136, 138. The sanctity of the cowpen, which appears in the earlie of the Vedas long before the sacred cow concept was formulated, might we reflect more widespread non-Aryan and perhaps pre-Aryan cattle cults in Indi Allchin, for example, suggests that the Neolithic ashmounds of the Decca resulted from the ritual burning of the dung in cowpens, just as in variou ceremonies in modern India, dung objects are consigned to fire and lamps are l in cowpens and stables (F. R. Allchin, *Neolithic Cattle-Keepers of South Indi* [Cambridge: University of Cambridge Press, 1963], p. 176).

40. Op. cit., p. 39.

41. For a general discussion of ahiṃsā in Indian philosophical thought, se Bashishtha Narayan Sinha, "Development of Ahiṃsā in the Vedic Tradition, *Prajna* 13, no. 2 (March, 1968):145–58; George Kotturan, *Ahiṃsā: Gautama t Gandhi* (New Delhi: Sterling Publishers, 1973); Kosheyla Walli, *Ahiṃsā i Indian Thought* (Varanasi: Bharata Manisha, 1974); and Unto Tähtinen, *Ahiṃsā Non-Violence in Indian Tradition* (London: Rider, 1976). The impact of ahiṃs on attitudes toward animals is examined in Burch H. Schneider, "The Doctrine c Ahiṃsā and Cattle Breeding in India," *Scientific Monthly* 67 (1948):87–92; an Gary S. Dunbar, "Ahimsa and Shikar: Conflicting Attitudes towards Wildlife i India," *Landscape* 19, no. 1 (1970):24–27.

42. F. Max Müller in *SBE* 1:1xxxvi.

43. *Chāndogya Upaniṣad*, 3.17.4.

44. *The Religion of India*, p. 198.

45. O. L. Chavarria-Aguilar, ed., *Traditional India* (Englewood Cliffs, Nev Jersey: Prentice-Hall, Inc., 1964), p. 23.

46. Dines Chandra Sircar, "Early History of Vaiṣnavism," in H. Bhattacharyya

ed., *The Cultural Heritage of India* (Calcutta: Ramakrishna Mission, Institute of Culture, 1956), 4:135.

47. Wm. Theodore De Bary, ed., *Sources of Indian Tradition* (New York: Columbia University Press, 1966), 1:35.

48. Keith, op. cit., p. 502.

49. A. S. Geden, "Upanishads," in Hastings, ed., op, cit., 12:540.

50. Basham, op. cit., p. 247.

51. For a detailed discussion of this, see Brown, op. cit., p. 36ff.

52. The text of Pillar Edict V is as follows:

> Thus saith King Priyadarśin, Beloved of the gods—When I had been consecrated twenty-six years the following animals were dẹclared unworthy of slaughter, namely parrots, starlings, ruddy geese, swans, *Nandimukhas*, *Gelatas*, flying-foxes, queen-ants, female tortoises, boneless fish, *Vedaveyakas*, *Ganga-Paputakas*, skates, tortoises and porcupines, hare-like squirrels, twelve-antler stags, bulls set free, household vermin, rhinoceros, grey doves, village pigeons, and all quadrupeds which are neither used nor eaten. She goats, ewes, and sows, which are with young or in milk, are unworthy of slaughter, and some of their young ones up to six months of age. Cocks shall not be caponed. Chaff containing living things shall not be burnt. Forests shall not be set on fire either for mischief or for the destruction of life. The living shall not be fed with the living. About the full moon of each of the three seasons and the full moon of Tishya, fish may neither be killed nor sold during three days, namely the fourteenth [and] fifteenth [of the fortnight] and the first [of the following fortnight], and certainly not on fast days. On the same days these and other species of life also shall not be killed in the elephant forest and fish preserves. On the eighth of [each] fortnight and on the fourteenth and fifteenth, on the Tishya and Punarvasu days, on the full-moon days of the three seasons,—on [such] auspicious days, bulls shall not be castrated: he-goats, rams, boars and such others as are castrated shall not be castrated. On the Tishya and Punarvasu days, on the full-moon days of the seasons, and during the fortnights connected with the full-moons of the seasons, the branding of horses and oxen shall not be done. Twenty-five jail deliveries have been effected by me, who am consecrated twenty-six years just in that period.

(D.R. Bhandarkar, *Aśoka* [Calcutta: University of Calcutta, 1955], p. 309.)

53. The freeing of bulls, a practice apparently of considerable antiquity, is still found in India today. The animals, usually dedicated to Shiva, are branded with his *triśul* and allowed to roam the streets unmolested.

54. N. A. Nikam and Richard Mckeon, eds., *The Edicts of Asoka* (Chicago: University of Chicago Press, 1959), p. 64.

55. V. A. Smith, *Asoka* (Oxford: The Clarendon Press, 1920), p. 66.

56. H. H. Sir Bhagvat Sinh Jee, *A Short History of Aryan Medical Science* (London: Macmillan and Co., 1896), p. 189. The emergence of the veterinary sciences, no doubt paralleling the medical sciences, occurs quite early in the history of civilization. There is evidence in cuneiform sources of the existence of veterinarians (translated literally as "physician of cattle"), and the Hammurabic Code of the eighteenth century B.C. sets out fees and liabilities for surgery performed on domestic animals.

57. *Arthaśāstra* 2.26.

58. Parpola et al., op. cit., p. 21.

59. Mrs. Sinclair Stevenson, *The Heart of Jainism* (London: Oxford University Press, 1915), p. 49.

60. The *Arthaśāstra* is attributed to Kauṭilya, the chief minister of Chandragupta Maurya, Ashoka's grandfather. Basham (op. cit. p. 79) suggests that it is a later, though pre-Guptan, work based on a Mauryan original rather than an authentic fourth century B.C. text.

61. 2.26.

62. 2.19.

63. 2.26.

64. 3.10.

65. 2.29.

66. The first mention of the word "goshala" that I have been able to detect occurs several centuries before the *Arthaśāstra* but uses it in the general sense of cowshed. Goshāla Mankhaliputta, leader of the Ājīvika sect and contemporary and rival of Mahāvīra during the sixth century B.C. received his name because he was born in a cowshed (*Bhagavatī Sūtra* 25.1).

67. James Legge, *A Record of Buddhistic Kingdoms* (New York: Dover Publications, Inc., 1965), p. 79. Other foreign accounts of India include John W. McCrindle, *Ancient India as Described by Megasthenes and Arrian* (London: Trübner, 1877); Flavius Arrianus, *Arrian, History of Alexander and India*, 2 vols. (Cambridge: Harvard University Press, 1949); Samuel Beal, trans., *Si-yu-ki. Buddhist Records of the Western World* (London: Kegan Paul, Trench, Trübner & Co. Ltd., 1884); J. Barthélemy Saint-Hilaire, *Hiouen-Thsang in India*, trans. from the French by Laura Enser (Calcutta: Susil Gupta, 1952); Thomas Watters,

trans., *On Yüan Chwang's Travels in India*, 2 vols. (London: Royal Asiatic Society, 1904–1905); Al Biruni, *Alberuni's India*, 2 vols., trans. E. C. Sachau (London: Trübner, 1910); Ibn Batuta, *Ibn Batuta's Travels*, 3 vols., trans. H. A. R. Gibb (London: Hakluyt Society, 1929–1962).

68. R. C. Sharma, "Animal Figures in Mathura Art," *Silver Jubilee Souvenir*, Uttar Pradesh College of Veterinary Science and Animal Husbandry, Mathura, n.d.

69. K. V. Subrahmanya Aiyer, "Tinnevelly Inscriptions of Maravarman Sundara-Pandya II," *Epigraphia Indica* 24, no. 22 (1937–1938):159.

70. An inscription dated A.D. 883–884 records the gift of a *gōsāsa*, a word that has yet to be explained but that may relate to goshalas. Fleet interprets this to mean cowshed, being a corruption of the Sanskrit goṣṭha, but Lakshminarayan Rao argues that it is probably an abbreviation or Kanarese rendering of *gōsahasra*, literally "gift of a thousand cows" (N. Lakshminarayan Rao, "Two Stone Inscriptions of Krishna II; Saka 805," *Epigraphia Indica* 21, no. 35 [1931–1932]:207).

71. "Hail! Fortune!

"When Jupiter stood in [the sign] Simha, in the year [denoted by the chronogram] Cholapriya [i.e., 1296], the prince Sarvanganatha, possessed of good report, from faith and to secure fame in abundance and for the sake of religion, reverentially built at the town of Syanandura a cow-house [goshala], a house of beautiful lamps, [and] Ah! an abode of Krishna, an open hall." (F. Kielhorn, "Three Inscriptions from Travancore," *Epigraphia Indica* 4, no. 27 [1896–1897]:203).

72. Brown, op cit., p. 38.

73. Ibid., p. 39.

74. Beal, op. cit., p. 214.

75. L. L. Sundara Ram, *Cow Protection in India* (George Town, Madras: The South Indian Humanitarian League, 1927), p. 177.

76. V. A. Smith, *The Early History of India* (Oxford: The Clarendon Press, 1924), p. 190.

77. François Bernier, *Travels in the Moghul Empire*, trans. A. Constable, ed. V. A. Smith (London: Oxford University Press, 1934), p. 236.

78. Quoted in Sundara Ram, op. cit., p. 184.

79. Ibid., p. 181.

80. Mukandi Lal, "Cow Cult in India," in A. B. Shah, ed., *Cow-Slaughter: Horns of a Dilemma* (Bombay: Lalvani Publishing House, 1967), p. 32.

81. *Shiva Digvijaya*, quoted in Sundara Ram, op. cit., p. 192.

82. Sundara Ram, op. cit., p. 191.

83. *Chāndogya Upaniṣad* 3.17.6.

84. H. C. Raychaudhuri, *Materials for the Study of the Early History of the Vaishnava Sect* (Calcutta: University of Calcutta, 1936), p. 39.

85. Basham, op. cit., p. 305.

86. *Ṛg Veda* 1.22.18; 10.19.14. *Baudhāyana Dharma Sūtra* 2.5.24.

87. Basham, op. cit., p. 305.

88. Bimanbehari Majumdar, *Kṛṣṇa in History and Legend* (Calcutta: University of Calcutta, 1969), p. 53ff.

89. Dines Chandra Sircar, op. cit., 4:123.

90. *Harivaṃśa Purāṇa* 2.16.2.

91. Though Goverdhan Pūjā and Gopāṣṭamī are among the more important of the cattle festivals of the pastoral peoples of India, there exists a great variety of other cattle-related festivals and ceremonies such as the Gāydāṇṛ festival of Bihar and Orissa (see Kalipada Mitra, "The Gāydāṇṛ Festival and Its Parallels," *Indian Antiquary* 60 [October 1931]:187–190 and 61 [December 1931]:235–238) suggesting an ancient and widespread cattle cult in India out of which, perhaps, the pastoral Krishna might have evolved.

92. Suvira Jaiswal, *The Origin and Development of Vaiṣṇavism* (New Delhi: Munshi Ram Manohar Lal, 1967), p. 151.

93. *Laws of Manu* 10.23.

94. See p. 156 of text.

95. Samuel Purchas, *Hakluytus Posthumus, or, Purchas His Pilgrimes, Contayning a History of the World in Sea Voyages and Lande Travells by Englishmen and Others*, 20 vols. (reprint of the 1625 ed.; New York: Macmillan, 1905–1907), 10:182.

96. John Ovington, *A Voyage to Suratt in the Year 1689*, ed. H. G. Rawlinson London: Oxford University Press, 1929), p. 177.

97. John Fryer, *A New Account of East India and Persia, Being Nine Years' 'ravels 1672–1681*, 3 vols., ed. William Crooke (London: The Hakluyt Society, econd Series, 1909–1915), 1:138.

98. William Crooke, *Things Indian* (London: John Murray, 1906), p. 371. 'avernier's actual description of Mathura, however, leaves this interpretation pen to question (Jean Baptiste Tavernier, *Travels in India*, 2 vols., trans. V. Ball, d. William Crooke [London: Oxford University Press, 2d ed., 1925], 1:63).

99. Niccolo Manucci, *Storio do Mogor: or Mogul India*, 4 vols., trans. W. rvine (London: John Murray, 1906–1908), 1:156; Peter Mundy, *The Travels of 'eter Mundy*, 5 vols., ed. R. C. Temple (London: The Hakluyt Society, 1907–936), 2:310; Jean de Thévenot, *The Travels of Monsier de Thévenot into the evant*, trans. A. Lovell (London: H. Clark, 1971), p. 13. See also Jan Huyghen van inschoten, *The Voyage of John Huyghen van Linschoten to the East Indies*, 2 ols., ed. Burnell and Tiele (London: The Hakluyt Society, 1885), 1:253; Pietro ella Valle, *The Travels of Pietro della Valle in India*, 2 vols., ed. E. Grey London: The Hakluyt Society, 1892), 1:67, 70; James Forbes, *Oriental Memoirs*, vols. (London: Richard Bentley, 1834), 1:156; Walter Hamilton, *A Geographi-al, Statistical and Historical Description of Hindoostan and Adjacent Countries*, vols. (London: John Murray, 1820), 1:718.

100. Louis Rousselet, *L'Inde des Rajahs* (Paris: Libraire Hachette et Cie, 875), p. 17.

V: Compassion for Life—Case Studies of Pinjrapoles in Modern India

1. The Ahmedabad Pinjrapole was visited by me on several occasions during ny stay in Ahmedabad between February and August, 1975. Information was nade available by permission of the Managing Trustee of the Society.

2. Sultan Ahmad Shah of Gujarat, founder of Ahmedabad and one of the four hmads in honor of whom the city was named.

3. Bombay (Presidency), *Gazetteer of the Bombay Presidency*, vol. 4, *Ahmeda-ad* (Bombay: Government Central Press, 1879), p. 250.

4. Ibid., p. 87.

5. Kenneth L. Gillion, *Ahmedabad: A Study in Indian Urban History* (Berke-ley and Los Angeles: University of California Press, 1968), p. 19.

6. Op. cit., p. 107.

7. Pers. Comm., Śri Hiralal Bhagwati.

8. Gillion, op. cit., p. 96.

9. Samuel Purchas, *Hakluytus Posthumus; or, Purchas His Pilgrimes, Contayning a History of the World in Sea Voyages and Lande Travells by Englishmen and Others*, 20 vols. (reprint of the 1625 ed.; New York: Macmillan, 1905–1907), 10:170.

10. Śravak is a term applied to the Jaina community and *bania* (vania) refers to the merchant classes, the *b* and *v* often being interchangeable in Hindi and related languages.

11. Peishwa, literally "Prime Minister," refers here to the period of Maratha rule. The title itself was applied to the chief minister of Shivaji and his successors. The Guicowars (Gaekwar) were the ruling house of Baroda and one of the most powerful Hindu princely families in western India.

12. *Petitions,* 9th July, 1827. General Department 13/146/1827. (unpublished records.) Maharashtra State Record Office, Bombay.

13. *Petitions*, 8th October, 1827. General Department 170/1828. (unpublished records.) Maharashtra State Record Office, Bombay.

14. *Report on Town Duties*, by J. Langford. Revenue Department 106/1069 1839. (unpublished records.) Maharashtra State Record Office, Bombay.

15. Op. cit., p. 112.

16. It was customary at this time for the management of the pinjrapole to lie in the hands of the Nagarseth.

17. Op. cit., p. 115.

18. The following mahājans are at present represented on the advisory board:

Maskati Cloth Market Association
Pañcakua Cloth Mahajan

Ahmedabad Cotton Brokers Association
Ahmedabad Cotton Merchants Association
Rani's Hajra Cloth Mahajan
Jhaveri Kanti Mahajan
New Madhupura Mahajan
Old Madhupura Mahajan
West India Cotton Association
Ahmedabad Grain Merchants Association
Ahmedabad Seeds Merchants Association
Manekchowk New Silk Cloth Mahajan
Astodia New Rangoti Cloth Mahajan
Bankers Mahajan

As all of these mahājans are trade guilds rather than caste associations, their membership is mixed but many of them are traditionally Jain and Jains often form the dominant element.

19. The levying of a tax for charitable purposes is a widespread practice among the mahājans of western India. In the *Rules and Regulations* (p. 6) of Ahmedabad's Pañcakua Cloth Mahajan, for example, it is set out that a lag of sixteen paisaa be paid by members on every Rs. 1,000 worth of sales and that of the total collected 50 percent should go to the Ahmedabad Pinjrapole, 20 percent to the mahājan's Educational Fund, 20 percent to other charitable funds, and the remaining 10 percent to pinjrapoles outside the city.

20. This rises to 30.68 percent if the government subsidy is excluded from the regular sources of income.

21. In addition to the great schism into Digambara and Swetambara, further subdivision into numerous sects and subsects has occurred in Jainism. Swetambara Jains are divided into Sthanakvasi and Deravasi (sometimes known as Murti Pujak, or "Idol Worshippers"), being distinguished from each other by the fact that the former do not worship idols. These divisions are essentially doctrinal, but the situation is further complicated by the existence of social distinctions based on caste. The chief castes among the Jains are the Oswal, Poravad, Śri Śrimali and Śrimali, each of which is further divided into Dasa and Visa, these being exogamous groups—hence the "Ahmedabad Jain Swetambara Murti Pujak Visa Śrimali Friendly Society" (Stevenson, in Hastings, ed., *Encyclopaedia of Religion and Ethics* 12:123).

22. In 1863, for example, the Surat Bakr Īd Fund was established by "certain charitably disposed Parsees and Hindoos in Surat and its vicinity with the intention and for the purpose of redeeming from slaughter all or some of the animals intended for sacrifice annually at Surat on the said Holiday of Bakaree-Id or Id-e-Kurbanee" (*Trust Deed* of the Bombay Pinjrapole dated 10th June

1871). Instrumental in the setting up of this fund was the Parsee philanthropist Sir Jamsetjee Jeejeebhoy, one of the first Trustees of the Bombay Pinjrapole

23. The true meaning of the word *upashray* is not accurately conveyed b either "temple" or "monastery." It is a place of worship for Sthanakvasi Jains, bu as sthanakvasis do not worship images or idols, the upashray is more of meeting hall rather than a temple. Jain monks or nuns often reside in th upashray, yet neither is it a monastery in the formal sense. This particula upashray is located in Pipardi Pol, one of the *pols* or neighborhoods o Ahmedabad. The pol is a unique feature of urban morphology described by th Gazetteer of the Bombay Presidency as a "house group, having only one or a most two entrances, protected by a gateway closed at night as a safeguar against thieves. Inside is one main street, with crooked lanes branching o either side. Most vary in size from five or ten to fifty or sixty houses. One o them . . . includes several smaller pols, with an area of about fifty acres and population of 10,000 souls." (op. cit., p. 294.)

24. As noted earlier, the entire village of Rancharda was donated to th pinjrapole by a former Nagarseth of Ahmedabad, while one Hathising Kesrising a member of the same family, endowed the institution with half the village o Mankol. Another patron of the pinjrapole was the Gaekwar of Baroda, the loca Maratha ruler. In 1862 the trustees of the pinjrapole approached the Gaekwar requesting land for the institution; this request being granted, grazing lands between the villages of Vanaya, Vansajda and Unali were "donated . . . to graz the cattle of Khoda Dhor (pinjrapole) . . . (this land) given in charity will b continued ever by the government." This memorandum from the Gaekwar's Administrative Office, dated 7 September 1862, is still in the possession of th Ahmedabad Pinjrapole Society.

25. Bombay (Presidency), *Gazetteer of the Bombay Presidency*, vol. 8, *Kathi awar* (Bombay: Government Central Press, 1884), p. 2.

26. S. R. Rao, *Lothal and the Indus Civilization* (New York: Asia Publishing House, 1974).

27. 300,614 according to the 1971 Census of India.

28. The 1891 Census of India shows Rajkot and its Civil Station having population of 29,247 of which 2,391 (8.18 percent) were Jains by religio (Census of India, 1891. *General Tables for British Provinces and Feudator States*, Vol. I).

29. The total animal population of the pinjrapole in March 1975 was 22 useless cows, 289 useless bullocks, 14 working bullocks, 10 female calves, 4 male calves, 22 useless buffalo, 101 goats, 8 horses, 4 sheep, 15 dogs, 1 cat, hens, 2 peacocks, 2 rabbits and approximately 700 pigeons.

VI: The Gift of a Cow—Case Studies of Goshalas in Modern India

1. The three institutions discussed here were all personally visited by me, the Dibrugarh goshala in May 1974, Vrindavan in November of the same year, and Nathdwara on several occasions in February 1975. Unless otherwise stated, statistics presented in the text were gathered at the time of these visits. Data were made available by courtesy of the Chief Executive Officer of the Temple Board of the Śri Nathji Mandir, and the Trustees and Managing Committees of the Vrindavan and Dibrugarh institutions.

2. Lt. Col. James Tod, *Annals and Antiquities of Rajasthan*, 3 vols. (London: Oxford University Press, 1920), 1:449.

3. Rajputana (Province), *The Rajputana Gazetteer* (Simla: Government Central Branch Press, 1880), 3:54.

4. Krishnalal Mohanlal Jhaveri, trans., *Imperial Farmans* (A.D. *1577 to* A.D. *1805) Granted to the Ancestors of His Holiness the Tikayat Maharaj* (Bombay: The New Printing Press, n.d.), Farman III A.

5. Ibid., Farman IV.

6. Animal fighting is an ancient pastime among the peoples of India. The *Tevigga Sutta* (late fourth century–early third century B.C.) forbids Buddhist monks witnessing "combats between elephants, horses, buffaloes, bulls, goats, rams, cocks and quail," (*SBE* 11:192). Fábri's evidence from the Indus Valley, although pitting man against animal rather than animal against animal, suggests that the tradition of animal sports in India is much older than even this ("The Cretan Bull-Grappling Sports and the Bull Sacrifice in the Indus Valley Civilization," *Annual Report of the Archaeological Survey of India 1934*–35, p. 96). Crooke notes fights between elephants, quails, partridges, rams, deer, buffaloes, and cocks during the British period (*Things Indian*, p. 11), and even today animal fights occur. This author has witnessed a fight between rams in a village in Uttar Pradesh, and cockfighting, though officially frowned upon, is apparently common in many parts of the country. Although animal fighting in modern times is held mainly for sport or amusement, a consideration of origins raises some interesting questions. Laufer, for example, argues that cockfighting is a modern survival of an ancient form of divination ("Methods in the Study of Domestications," *Scientific Monthly* 25 [1927]:255). Along similar lines, one wonders whether the setting of cattle on a pig in the Gāydāṇṛ festival of eastern India might not represent a symbolic confrontation between opposing cultures.

7. An unusual feature of the Nathdwara goshala is the staging of buffalo

fights at the time of Daśehra, for which the institution keeps seven buffalo bulls. Daśehra (Durgā Pūjā) is the ten-day festival, beginning on the first day of Asvi (September–October), held in honor of Durgā, the war-goddess of the Hindu pantheon. In eastern India—where Durgā is worshipped in the form of Kālī, the goddess of death—and in Nepal, Daśehra is widely celebrated by the sacrifice of animals, especially buffalo—symbolic of Durgā's slaying of the evil buffalo demon Mahesha.

8. Uttar Pradesh (State), *Uttar Pradesh District Gazetters, vol. 36, Mathura* (Lucknow: Superintendent of Printing and Stationery, 1968), p. 16.

9. The handprint is popularly considered to be a charm against the evil eye.

10. For a discussion of the swastika, its origins and diffusion, see Thomas Wilson, *The Swastika, the Earliest Known Symbol, and Its Migrations; With Observations on the Migration of Certain Industries in Prehistoric Times* (Washington, D.C.: Government Printing Office, 1896).

11. When this matter was raised wih the goshala management, it was explained that the best milker in the institution was chosen for the gopūjā ceremony and that this was usually a western cross. A small point, perhaps, but significant in a culture that tends to view zebu and common cattle as breeds apart.

12. The name Jhunjhunuwala itself translates literally as "a person from Jhunjhunu," a town some 200 kilometers southwest of Delhi in the Jhunjhunu District of Rajasthan.

13. It seems to be the custom all along the Jumna and Ganges to dispose of carcasses by placing them in the river. Some informants claim that this practice is related to the sanctity of the rivers, while the more cynical suggest they form a convenient garbage dump—in either case, the residual value of the carcass is lost.

14. This is not uncommon in India, where the majority of farmers cannot afford to purchase or maintain costly mechanized equipment. Yet the tractor is much sought after, since not only is it much faster than traditional methods, but it also makes deeper furrows than bullocks pulling wooden ploughs. Many cooperatives, agricultural organizations, and even private individuals make tractors available for hire. The renting of its tractor earned the Vrindavan institution Rs. 19,769 profit for the year in question.

15. These include tea, rice, jute, rape, mustard, and sugar cane. Tea processing and oil milling are important food-processing industries in the area.

16. 11,227 according to the 1901 Census of India.

17. Vol. 11, p. 343.

VII: Goshalas and Pinjrapoles in Modern India—Organization, Management, Community Support, and Finances

1. This does not, of course, apply to institutions such as the Nasik Pinjrapole, which, as a subcenter of the Bharatiya Agro-Industries Foundation, Uruli Kanchan, has become a modern cattle-breeding farm working with artificial insemination techniques, crossbreeding, and cattle development in general.

2. The Chhapriali Pinjrapole near Palitana, for example, is run by Anandji Kalyanji, a Jain religious trust with its head office in Ahmedabad. The Idar Pinjrapole, north of Ahmedabad, was founded by Vijay Kamalsuri Sorji Maharaj, a Jain *muni* of the Swetambara sect.

3. Although Hinduism itself is sometimes designated as Sanātana Dharma, the Eternal Dharma, the term also appears to be applied to those who follow Brahmanical Hinduism as opposed to belonging to any specific sect.

4. The Arya Samaj is a modern Hindu reform movement founded in 1875 by Swami Vivkand Dayananda. Claiming to purge Hinduism of all the degenerate features that have appeared since the Vedic period, the Arya Samaj harks back to the Vedas as the source of all knowledge and truth, and as the foundation on which Hinduism should be rebuilt.

5. King Agarsen, so the story goes, had no sons. He consulted various gūrūs on this matter, and they advised him that he should save the cows. This he did and he was rewarded by fathering eighteen sons. As a result, he directed his descendants (Agarwals) to preserve cows throughout the generations; hence the creation of the goshala and the Agarwal concern for cow protection.

6. For example, Akbar's ban on cow slaughter, the farmans being preserved in a series of inscriptions at the Adiswara Temple near Palitana.

7. The gentleman concerned, one Mohammed Sakadat Siddiki, explained his association with the goshala in terms of social service, the desire to protect animal life, and an interest in developing resources rather than in religious

terms. The situation in Katni differs significantly from that in other goshalas where strong feelings were expressed about even accepting contributions from Moslems.

8. Some popular Buddhist views concerning specific food avoidances are worthy of mention. Tibetan refugees interviewed in Darjeeling commonly believe that fish, pigs, and hens are sinful creatures because they are carnivores (presumably scavengers) and that they are not eaten because their sins would pass on to their consumers. However, a lama who asserted he had searched the Buddhist literature for the specific origins of this belief could find no indication of these animals being regarded as especially sinful.

9. That this practice is still followed today can be attested to by me, having witnessed (in the Himalayas) the purchase of animals by Tibetan Buddhists to prevent them from being eaten.

10. In Darjeeling, for example, a lama makes regular visits to Siliguri to release fish into the Tista River on behalf of devout Buddhists who desire to gain merit. On one specific occasion, this was done for a Tibetan woman who wished to eliminate sin in the name of her long dead mother. She purchased Rs. 1,099 worth of fish in Siliguri—privately, since the local fish-eating Bengalis often object to the fish being taken off the market. The fish were transported down stream so they would not be recaptured after release, blessed by the lama and released into the river. Similar practices may be seen in Southeast Asia. Outside the Temple of the Jade Buddha in Bangkok, street vendors sell birds to visitors who then release them as a means of acquiring merit.

11. Buddhism has been the major vehicle in the spread of certain ideas influencing attitudes toward animals from India into the rest of Asia. Although goshalas and pinjrapoles are uniquely Indian, there are scattered references to institutions that serve similar functions in China. The Temple of the White Clouds (*Pai-yüan kuan*) in Peking, for example, used to maintain a shed for old pigs and sheep (*Hsin-sheng pao*, Taipei, 22 February 1976). A late nineteenth century source, *Sung-nan meng ching lu*, tells of a man who bought old cows and kept them in his "Office for Given-up Cows" (*Fang-niu chü*), "given-up" in the sense of "let loose." Later he added a "dog tent" (*kou-p'eng*). These and other practices such as the freeing of animals to gain religious merit reflect a reverence for life and belief in ahiṃsā that originate in India. Of specific interest are prohibitions against cow slaughter and the eating of beef in Asia which, while being associated with Buddhist beliefs, may perhaps ultimately be traced to Hindu attitudes toward the cow. Since Tang times (seventh–tenth century), laws have existed banning the slaughter of cows (A. F. P. Hulsewe, *Periodieke executie- en Slacht-Verboden in de T'ang tijd en Hun oorsprong* (Leiden: E. J. Brill, 1948]. In the Manchu period (1644–1911), the slaughter of cows and horses was forbidden under heavy punishment (G. Boulais, *Manuel du Code Chinois* [Chang-Hai: Imprimerie de la Mission Catholique, 1923]). I am indebted

to Professor Wolfram Eberhard of the University of California, Berkeley, for bringing these references to my attention.

12. Frederick J. Simoons, *Eat Not This Flesh* (Madison, Wisconsin: University of Wisconsin Press, 1961), p. 62.

13. Langford reports that the Broach Pinjrapole, in a deed dated 1793, levied the following collections on all imports and exports passing through Broach:

Goods		Tax (Rupees)	
Imports			
1. Indigo	per Maund	0-8-0	By sea or land
2. Grain of all sorts	per Culsey		1 seer by sea
3. Grain of all sorts	per Cart		2½ seers by land
4. Jinjeely seed	per Maund		1 seer by sea
5. Castor oil seed	per Maund		1 seer by sea
6. Sorungee root	per Maund		a malwa
7. Jinjeely seeds	per Cart		2½ seers by land
8. Castor oil seeds	per Cart		2½ seers by land
9. Coarse piece goods	per Bale	0-8-0	on transit through Broach
Exports			
10. Cotton seeds	per Culsey	0-0-6	by sea
11. Piece goods of Broach manufacture	per 12 Corge	1-0-0	by sea or land
12. Ditto coarse coloured	per 12 Corge	0-2-0	by sea or land
13. Castor or jinjeely cakes	per Candy	0-1-0	by sea or land
14. Mowra	per Culsey	0-2-0	by sea
15. Cotton packed at the Broach screws	per Candy	0-2-0	by sea
16. Grain of all sorts	per Culsey	0-1-0	by sea
17. On every bale of cotton cloth	per Bale	0-4-0	manufacture of Broach
18. On sarees	per Corge	0-0-6	by land
19. Cotton seed	per Culsey	0-½-0	by land
20. On all insurance made at Broach	per Rs. 100	0-1-0	by sea
21. On all marriages		1-8-0	each
22. Ditto of Bania Shravaks		3-2-0	each

(J. Langford, *Report on Town Duties*, Revenue Department 106/1069/1839. (unpublished records.) Maharashtra State Record Office, Bombay, p. 52)

14. This deed is in the possession of Anandji Kalyanji in Ahmedabad.

15. In addition to outright grants of land, royal patronage of animal homes included leasing of grazing land at reduced rates, annual subsidies, and cash gifts on special occasions. Thus, the mahājans of Patan held the village of Khalipur and 1898 bighas of land from Baroda State on payment of an annual fee of Rs. 75 for the support of the animals in its pinjrapole. On the birth of a son to the Mahārāja of Bhavnagar on 19 May 1912, he ordered that in addition to proclaiming a holiday, releasing prisoners and firing an eleven-gun salute, "Rs. 300 shall be given in support of the Cow-houses at Bhavnagar and Mahuva and of the cows in the Vaishnav Haveli at Bhavnagar" (*Report on the Administration of Bhavnagar State for the Year 1911–1912*, p. 35). Similarly, the Administration Report for Morvi notes that ". . . the annual grant of Rs. 12,000 to the local Pinjrapole was continued . . ." (*Report on the Administration of Morvi State for the Year 1934–35*, p. 45). This support is not limited to Hindu rulers alone. The management of the Palanpur Pinjrapole reports that the Nawab of Palanpur, a Moslem, instituted in 1947 an annual grant to the pinjrapole of Rs. 8,000, a sum that continued to be received intermittently after the state merged with the Union of India.

VIII: Goshalas and Pinjrapoles—Religious Ideals, Functions, and Festivals

1. The following discussion of Jainism is based on Wm. Theodore De Bary, ed., *Sources of Indian Tradition*, 2 vols. (New York: Columbia University Press, 1966); James Hastings, ed., *Encyclopaedia of Religion and Ethics*, 13 vols. (New York: Charles Scribner's Sons, 1955); E. Washburn Hopkins, *Ethics of India* (New Haven: Yale University Press, 1924); S. Radhakrishnan, *Indian Philosophy*, 2 vols. (New York: The Macmillan Company, 1923); W. Schubring, *The Doctrine of the Jainas*, trans. Wolfgang Beurlen (Delhi: Motilal Banarsidass, 1962); and Mrs. Sinclair Stevenson, *The Heart of Jainism* (London: Oxford University Press, 1915).

2. Albert Schweitzer, *Indian Thought and Its Development* (London: Adam and Charles Black, 1956), p. 79.

3. It is, perhaps, significant that these were the main sacrificial animals of Vedic Brahmanism, against whose excesses both Jainism and Buddhism appear to be a reaction.

4. De Bary, op. cit., 1:206.

5. *Bhagavad Gītā* 18.44.

6. This view is not restricted to the Vaiśya castes alone. When Lewis questioned villagers in Punjab on their interpretation of the *dharmata*, or the

"righteous man," included in the list of characteristics were contributions to the goshala, and the collection of food and fodder for the goshala. These responses came mainly from the upper castes, Jats and Brahmans (*Village Life in Northern India* [Urbana, Illinois: University of Illinois Press, 1958], p. 258). Many Brahman informants of this author included gorakshan and support for the goshala in their definitions of dharma.

7. Parsee concern for dogs can be traced back to the ancient Zoroastrians, who regarded the dog as the most sacred of all animals associated with their god, Ahura-Mazda. Passages in the Zend-Avesta extoll the virtues of the dog and set out penalties for the mistreatment or killing of the animal (SBE 4:151 ff.). It is the custom among the Parsees to keep a dog in the presence of a dying man, muzzle near his mouth, so that it may receive his parting breath and guide the soul to heaven. Among Hindus, attitudes toward the animal are mixed. The dog is worshipped in connection with the cult of Bhairon, one of the lesser godlings of India, but many regard it as unclean. Crooke ascribes the replacing of the faithful hound by the mongoose in the popular version of the Bethgelert legend to Brahmanical prejudice against the dog (*The Popular Religion and Folk-Lore of Northern India,* 2 vols. [Westminster: Archibald Constable and Co., 1896] 2:219, and *Things Indian* [London: John Murray, 1906], p. 147).

8. At least one institution, the Pinjrapole Goshala at Jabbalpur reports celebrating Janamaṣṭamī by holding a procession of cows through the streets of the town.

9. Polā, also called *Bel Polā,* is widely celebrated throughout Maharashtra, although its date varies from district to district. The following account is based on data gathered at a Bel Polā held at Narayandoho in Ahmednagar District in 1974, the festival falling on the dark night (amas) of the month of Shravan.

Bel Polā is a day of rest for the bullocks of the village. In the morning, they are washed in the river and then taken home, where they are fed millet and molasses. In the afternoon, the owners of the bullocks decorate their animals. The horns are painted red with *hingul,* and wrapped with colored paper. The bodies are painted with designs and daubed with political slogans, the names of the animals, and even popular film titles. Strings made from the roots of trees are tied to the front hooves of the bullocks. Even male buffaloes are decorated, although they do not take part in the subsequent proceedings. All the working bullocks in the village are included, even those belonging to Moslems and *harijans.* In the early evening, the bullocks are led in pairs in a procession around the outskirts of the village and then brought in to the main square. There, the officiating Brahman commences the pūjā ceremonies by invoking Ganesh, represented by an areca nut. After chanting mantras and sprinkling the nut with water, *kum-kum* (a red powder) and dry rice, the pūjāri proceeds with gopūjā. A bullock belonging to the village headman is brought forward for the ceremony, the pūjāri sprinkling first water, then red powder, turmeric, and rice on its front right hoof, and afterward on its forehead (fig. 20). The pūjā ceremony complete, all the other bullocks in the procession are led in front of the one to which pūjā has just been made and forced to kneel before it. The villagers then disperse, taking their bullocks with them, to their own houses, where the women wait to perform pūjā to the animals. A circle is made around the head of each bullock with a *pūrī*, which is then thrown away, the object of this being to satisfy astral bodies and spirits and to distract them away from the bullocks. Red powder and turmeric are sprinkled on the foreheads of the animals, which are then fed *pūran pūrīs,* pūrīs containing gram and molasses. Grain is given to the bullocks, though a portion is kept to be placed with the seed stock for the following year. Finally, gold ornaments are waved in a wide circle around the heads of the animals to bring wealth and good fortune for the coming year. Festivities, with music and dancing, continue late into the evening. Though not related to Bel Polā, the following day, the 1st of Bhado, is also a festive occasion with goat sacrifice and a community feast, it being the custom not to eat meat during the month of Shravan.

10. "Baras" is the 12th day of the fortnight; "Vashu" means animal and "Vats" translates as child or dear one. In the Jodhpur region, Vats Baras is popularly referred to as the "Day of Calves and Children."

11. In Maharashtra, it is commonly believed that if a child is born on an inauspicious day, he should be wrapped in a cloth and placed before a cow at Vashu Baras. The child is then said to have been born of the cow's womb, and thus the stigma of his inauspicious birth is removed.

12. *Hari-Bhakti-Vilāsa* 6.32.

13. Ibid. 6. 74–79.

14. 6.81.

15. 6.80, 6.98.

16. 6.88–92.

17. 6.93–97.

18. *Bhavisya Purāṇa* 19.271; *Matsya Purāṇa* 19.274.

19. Other foods offered to the Deity include *śrikhand, kuliya, khīr, khirāra, rasala, jadava, sikharini,* and *abhiksa.*

20. Two other practices at the Jagannath Mandir Goshala are worthy of mention. In most goshalas, gopūjā is usually performed on special occasions such as Gopāṣṭamī or Janamaṣṭamī, but in Ahmedabad *gouri gopūjā* is a common daily occurrence. It is the custom in Gujarat for young unmarried girls (*gouri*) to take a

vow to perform gopūjā every day to seek the blessings of the gods at times of adversity, for obtaining a husband or for similar purposes. Thus, in keeping with their vow, young girls make daily visits to the temple goshala to perform the cow worship ceremony. The second practice, too, is a reflection of basic Hindu attitudes toward the cow. There are numerous rites and ceremonies in Hindu life which involve gopūjā or which center around the cow. A Purāṇic story, for example, tells how Krishna once saw a number of lost souls groping in the dark, unable to reach the Kingdom of Death. He was moved to pity, and so changed them into cowherds. By grasping the tails of their cows the transformed spirits crossed the Vaitarani River (the Hindu Styx) in safety and entered the realm of Yama. This event is still celebrated in the *Gai-jatra* festival in Nepal, and it is a popular belief among Hindus that the soul of a dying man will be helped across the River of Death by his touching the tail of a cow. If a cow is not available for such ceremonies, one may be requested from the Jagannath Mandir Goshala, which will send an animal to the home for pūjā or other rites.

21. Bombay (Presidency), *Gazetteer of the Bombay Presidency,* vol. 4, *Ahmedabad* (Bombay: Government Central Press, 1879), p. 113.

22. This practice is by no means unique to modern India. Much of the corpus of Indian inscriptions is devoted to recording gifts and donations made in the name of religion.

IX: Economic Activities, Functions, and Ecological Mechanisms

1. The following description of the institution in 1879 is provided by the Bombay Gazetteer:

> The immediate cause of the founding of the home was a police order to catch stray bulls and kill stray dogs. The Gujarati inhabitants of the city formed a committee and took charge of all stray cattle and dogs, and since then the home has become a permanent institution. All animals, healthy, maimed, diseased or old, are received, though the rule is to attend only to the disabled and unserviceable. Except to the poor, admission fees are charged at the rate of £2 10s [Rs. 25] on horses and 6s [Rs. 3] on oxen, cows and buffaloes. Birds are taken free of charge but any amount paid on their account is accepted. When necessary a Muhammadan farrier is called in to treat sick horses. The other animals are treated by the servants of the home. Healthy animals are given grass and the sick fed on pulse and oilcake. Healthy animals are made to work for the home. After recovery animals are given free of charge to any one who asks for them and is able to keep them. The home has two meadows or *kurans* near the city . . . the produce of these two meadows suffices for the wants of the home. In 1879 the home had about 200 head of cattle and 100 birds. In May, when most of the cattle and two deer were away at

the grazing grounds, the home had 10 horses, one *nilgáy*, a buck, and an antelope in a stable, about twenty peafowls in a square railed off at the end of the stable, three or four monkeys with running chains on a pole under a large tree, two foxes, a hare, two rabbits, and a number of pigeons, some fowls and a turkey. Besides these the home had one or two cows, a few goats, some bullocks and sheep. . . . The home is managed by a committee of whom in 1879 four were Hindus and two were Parsis. . . . The home has a yearly revenue of about £150 [Rs. 1,500] chiefly from cesses from groceries at 1½d [1 a.] a bag, on jewelry sales at ¼ per cent, on bills of exchange at 5/32 per cent, and on grain at 1/64 per cent (*Gazetteer of the Bombay Presidency*, vol. 18, *Poona*. Pt. 3, p. 332).

2. Of the twenty-five institutions listed in table 13, nine reported no income at all from the sale of dung. In one goshala, dung accounted for 12.85 percent of the annual income, but the average for the table is a mere 2.9 percent.

3. A survey of 1,062 goshalas and pinjrapoles undertaken by the Central Council of Gosamvardhana in 1954–1955 showed their total land resources to be 218,280 acres, of which 29.44 percent was cultivable and the rest was grazing land or under buildings (Harbans Singh, *Gaushalas and Pinjrapoles in India* [New Delhi: Central Council of Gosamvardhana, 1955] p. 20).

4. According to my sample survey, the land resources of goshalas and pinjrapoles amounted to 0.39 acres grazing and 0.12 acres cultivated land per head of cattle, compared to the Central Council of Gosamvardhana's 1954–1955 estimate of 1.17 acres grazing and 0.49 acres cultivable land. These are admittedly crude figures masking a range of variables, but they do provide some indication of the resources available. Cattle:land ratios today are being modified further by land-tenure reforms and legislation, in various states, establishing limits on land holdings.

5. Some 187 acres of the Siliguri Goshala land were expropriated in 1969 as a part of the then Communist State Government's land-reform program.

6. It is perhaps a significant comment on the attitudes of both government and the recipient organizations that most of these schemes are directed toward cattle rather than buffalo development.

7. This practice is reflected in the predominance of female calves over males in most goshalas and pinjrapoles.

8. Most of these are smaller institutions with cattle populations considerably less than a hundred. Thus, at the Śri Lakshmi Goshala in Saidpur, a small town on the north bank of the Ganges in eastern Uttar Pradesh, out of a herd of seventy-five cattle, only three cows were in milk and another seven with calf. The remaining animals were useless cows and bullocks. Similarly, in nearby Ghazipur, the Śri Gopal Krishna Goshala maintained only nineteen animals, with no milk cows

and only two between lactations. The minimal quantities of milk produced in institutions such as these are consumed by goshala workers or members of the goshala society.

9. The goshalas in Gokul, near Mathura and Azamgarh, north of Varanasi, report average daily yields for 1973 to be 0.7 kg. and 0.36 kg., respectively.

10. The Calcutta Pinjrapole Society's milk production for the year April 1973–March 1974 amounted to 327,637.61 kg. The milk is shipped by train from branches outside the city to the Society's Milk Depot in Calcutta itself, and then distributed in the city by the Society's milkmen.

11. Though the Indian farmer may favor Indian breeds over the exotic, some goshala managements appear to have accepted the superior milking qualities of western crosses since fifteen institutions in the sample survey have introduced western breeds. According to statistics kept by the Nasik Pinjrapole for their western/Indian crossbred cows, average yields per milk day for 25 percent Holstein-Friesian/75 percent Gir cows was 5.7 kg.; for 50 percent Holstein-Friesian/ 50 percent Gir, 6.6 kg.; and for 75 percent Holstein-Friesian/25 percent Gir, 7.3 kg. This compares with 4.2 kg. for purebred Gir cows kept at the institution.

12. Hoffpauir noted that though in 1961 the buffalo was the main milking animal in Delhi, Bombay, and Madras, the zebu predominates in Calcutta ("India's Other Bovine: A Cultural Geography of the Water Buffalo" [Ph.D. dissertation, University of Wisconsin, 1974], p. 128). This situation had changed by the early 1970s since the provisional figures of the 1972 Livestock Census show 12,121 buffaloes in milk compared to 6,852 zebu in the city limits. Local city ordinances restrict the keeping of dairy herds in Calcutta itself, although illegal milkmen are found throughout the city. The number of zebus in milk in neighboring districts (91,441 in Howrah and 54,076 in Hooghly) indicates that here lies the main source of milk for the city.

13. Shah notes that similar practices in Bombay during the early decades of this century resulted in the replacement of entire herds of milk buffalo every eight or nine months, with newly imported animals replacing those sent to the slaughterhouse after a single lactation because of the cost of feeding them during their dry periods or of returning them to rural districts (M. M. Shah, "Cow-Slaughter: The Economic Aspect," in A. B. Shah, ed., *Cow-Slaughter: Horns of a Dilemma* [Bombay: Lalvani Publishing House, 1967], p. 57).

14. The West Bengal Animal Slaughter Control Act of 1950 prohibited the slaughter of cattle and buffalo unless the animal were over fourteen years of age and unfit for work or breeding, or permanently incapacitated owing to age, injury, deformity, or disease. Although this applied to the entire state, a later Slaughter Control Act in 1966 restricted its jurisdiction to municipal areas only.

15. According to H. Guha, Director of Animal Husbandry for West Bengal, cattle culled from government farms between 1953 and 1955 were given to goshalas, complete with a donation of Rs. 20 per head. After that time, such animals were auctioned, many of them being purchased by cow-protection societies. Today, however, some 80–100 percent of the culled animals go to slaughterhouses.

X: Animal Homes in National Planning and Economic Development

1. India (Republic), *The First Five Year Plan* (New Delhi: Planning Commission, 1952), p. 7.

2. Ibid., p. 273.

3. Ibid, p. 274.

4. Sardar Datar Singh, *Reorganisation of Gaushalas and Pinjrapoles in India* (New Delhi: Ministry of Agriculture 1948), p. 8.

5. United Provinces, Ajmer-Merwara, Orissa, West Bengal, Central Provinces and Berar, Bihar, and Jaipur State.

6. Between 1945 and 1950, Goshala Development Schemes were started in Ajmer-Merwara (1945), Assam (1947), Bihar (1949), Bombay (1948), Central Provinces and Berar (1946), Orissa (1946), United Provinces (1946), and West Bengal (1945), and were under consideration in Madras and the Punjab.

7. These were Ajmer-Merwara (Federation founded in 1945), Bihar (1945), Bombay (1945), Central Provinces (1946), East Punjab and Delhi (1948), Jaipur State (1946), Orissa (1945), United Provinces (1947), Cutch (1949), West Bengal (1948), and Assam (1949).

8. The full text of the Act is presented in Appendix B.

9. Ministry of Food and Agriculture Resolution No. F 9-40/51-L dated 30 January 1952.

10. Harbans Singh, *Gaushalas and Pinjrapoles in India* (New Delhi: Central Council of Gosamvardhana, 1955).

11. The proceedings of these seminars, all published by the Central Council of Gosamvardhana, are found in *First Gosamvardhana Seminar (Srinagar)* (28

August–1 September 1958), *Second Gosamvardhana Seminar* (*Mount Abu*) (20–22 June 1960), *Gosamvardhana Seminar on 'Approach to Cattle Development in the Country'* (*Bombay*) (2–4 October 1963).

12. The Central Council of Gosamvardhana's funds were received in the form of an annual grant-in-aid from the Government of India.

13. *The First Five Year Plan,* p. 276.

14. These had been established in Bihar, Uttar Pradesh, Pepsu, Coorg, Bhopal, Kutch, Vindhya Pradesh, Tripura, and Saurashtra.

15. The estimated cost of setting up a gosadan with a capacity of two thousand cattle was Rs. 50,000, with subsequent recurring costs of Rs. 20,000 per annum.

16. India (Republic), Ministry of Food and Agriculture, *Report of the Expert Committee on the Prevention of Slaughter of Cattle in India* (Delhi: Manager of Publications, 1955), p. 62.

17. India (Republic), *Review of the First Five Year Plan* (New Delhi: Planning Commission, May 1957), p. 129.

18. India (Republic), *Second Five Year Plan* (New Delhi: Planning Commission, 1956), p. 283.

19. The Gosadan at Gularbhoj (Nainital District) in Uttar Pradesh was taken over by the Central Council in 1961. According to the Council's Annual Report for 1965–1966, the institution handled a total of 4,474 cattle during the year. After taming, 295 wild cattle were sold to breeders, 3,430 hides obtained, and 38,615 kg. of bone meal, 10,425 kg. of meat meal, and 97 kg. of fat were produced. Receipts for the year amounted to Rs. 86,488 versus expenditures of Rs. 74,829.

20. The Doms are regarded as one of the lowliest and most polluted castes in all of India, traditionally being scavengers and concerned in the removal of human corpses. They also gather wood for funerals and handle the cremation of corpses at the burning *ghats.*

21. *Gosamvardhana Seminar on 'Approach to Cattle Development in the Country,'* p. 48.

22. Sardar Datar Singh, op. cit., p. 9.

23. *Second Five Year Plan,* pp. 283–284.

24. *First Gosamvardhana Seminar,* p. 131.

25. India (Republic), Ministry of Food and Agriculture, Department of Agriculture, *Report of the Committee for the Prevention of Cruelty to Animals* (Delhi: Manager of Publications, 1957), p. 125.

26. *First Gosamvardhana Seminar,* pp. 123–145.

27. India (Republic), Central Council of Gosamvardhana, *Annual Report for the Year Ending 31st March, 1961,* p. 7.

28. India (Republic), Central Council of Gosamvardhana, *Annual Report for the Year Ending 31st March, 1963,* p. 39.

29. Ibid. p. 30.

30. India (Republic), Planning Commission, *Draft Fifth Five Year Plan, 1974–79* (Delhi: Manager of Publications, 1974).

31. In Haryana, for example, the state's fourth Five Year Plan allocated Rs. 245,000 for goshala development, yet none of this was ever utilized (Government of Haryana, Planning Department, *Fourth Five Year Plan 1969–74*, p. 105). Similarly, Gujarat planned to develop fifty institutions during its fourth plan, but reduction in funds owing to scarcity conditions drastically curtailed its program.

GLOSSARY

Agarwal: a Vaiśya trading caste, originally from Rajasthan

ahiṃsā: the philosophy of nonviolence to living creatures common to the Hindu, Buddhist, and Jain religions

ārtī: a Vaishnava ceremony involving the waving of *ghī* lamps in a circular motion before the head of the image of Krishna

Bakr Īd: a Moslem festival, celebrated on the 10th of Dhu'l Hadjdja, at which animals are sacrificed

bhakti: devotion to a personal god, especially as practiced by the followers of Krishna

bigha: a measure equivalent to five-eighths of an acre

darśan: a term used by Vaishnavas to refer to the coming of the worshipper into the presence of the image of the deity; Vallabhāchārya's teachings lay emphasis on the manifestation of god's grace through the sight of his image

dharma: the Hindu concept of duty or righteous conduct; though often used loosely in the sense of "religion," the word, which is interpreted slightly differently in Hinduism, Buddhism, and Jainism, has no precise translation in English

dharmada: charitable donations (of money, grain, land, etc.) made in the name of *dharma*

ghī: clarified butter, one of the five products of the cow

godān: the gift of a cow, formerly made to Brahmans for services rendered

Gopāṣṭamī: a festival commemorating the occasion when Krishna first took his father's herds of cattle to graze in the forests of Vraj

gopūjā: the cow-worship ceremony

gorakshan: cow protection

gosadan: Government-sponsored camps for concentrating stray and useless cattle in isolated rural areas

gosamvardhana: cow protection and development

Gosamvardhana: a journal on cattle development and related topics, published by the Central Council of Gosamvardhana

goshala: a charitable institution dedicated to preserving sick, disabled, and otherwise useless cattle

goṣṭha: a cowpen or pasture for cattle (Sanskrit)

Goverdhan Pūjā: a festival commemorating Krishna's defiance of Indra, celebrated in villages, temples, and *goshalas*

Janamaṣṭamī: Krishna's birthday

jīv (jīva): the Jain concept of "life" or "soul"

jīvat khān: rooms maintained by *pinjrapoles* for the preservation of insects, vermin and other such life forms

jīv-dayā: the Jain concept of "compassion for life"

kabūtriya: large, community-supported birdhouses found in towns and villages in Gujarat

karma: the belief that deeds performed in the past influence one's present life and the nature of future incarnations

khuraki: a system whereby cattle may be boarded in animal homes

Kṣatriya: the second (warrior) *varna* or class into which Aryan society was divided

kutta-kī-rotī: funds maintained by *pinjrapoles* for feeding dogs

lag: a tax paid by merchants and traders for the support of animal homes and other charitable institutions

mahājan: a trade guild or business association in western India, traditionally associated with the organization and financing of *pinjrapoles*

maund: a unit of weight equivalent to 37.39 kg. or 82.28 lb.

pañcaṃrit: "nectar" made from the products of the cow and used in Vaishnava ritual

pañcagavya: the five products of the cow: milk, curds, clarified butter, urine, and dung

pārabaḍī: a birdhouse found in Gujarat and western India, where grain is left out to feed birds

pinjrapole: a Jain institution found mainly in western India, dedicated to the preservation of animal life

prasād: offerings to the deity, usually sweetmeats made from milk products

pūjā: worship

Pushti Marg: literally "The Way of Grace," the sect of Krishna worshippers who follow the teachings of Vallabhāchārya

śakti: the active, energetic aspect of a god personified as his wife, and usually associated in western India with the worship of various forms of the Mother Goddess

Vaiśya: the third (mercantile) *varna* or class into which Aryan society was divided

vania (*bania*): a member of the merchant or trading castes

BIBLIOGRAPHY

Abbott, John. *The Keys of Power: A Study of Indian Ritual and Belief.* Secaucus, New Jersey: University Books, 1974.

Abul Fazl. *Ain-i-Akbari.* 3 vols. Translated by H. F. Blockman and H. S. Jarrett. Calcutta: Royal Asiatic Society of Bengal, 1873–1896.

Adams, Richard Newbold. *Energy and Structure: A Theory of Social Power.* Austin: University of Texas Press, 1975.

Al-Biruni. *Alberuni's India.* 2 vols. Edited by Edward C. Sachau. London: Kegan Paul, Trench, Trübner and Co., 1910.

Albright, W. F., and P. E. Dumont. "A Parallel between Indic and Babylonian Sacrificial Ritual," *Journal of the American Oriental Society* 54 (1934):107–128.

Allchin, F. R. *Neolithic Cattle-Keepers of South India.* Cambridge: University of Cambridge Press, 1963.

———. "Early Domestic Animals in India and Pakistan," pp. 317–322 in the *Domestication and Exploitation of Plants and Animals.* Edited by Peter J. Ucko and G. W. Dimbleby. Chicago: Aldine Publishing Co., 1969.

———, and Bridget Allchin. *The Birth of Indian Civilisation.* London: Penguin Books, 1969.

Ambedkar, B. R. *The Untouchables.* New Delhi: Amrit Book Company, 1948.

Anand, Roshan Lal. *The Milk Supply of Lahore in 1930.* Board of Economic Inquiry, Panjab. Rural Section Publication no. 28, 1933.

Anstey, Vera. *The Economic Development of India.* London: Longmans, Green and Co., 1936.

Arrianus, Flavius. *Arrian, History of Alexander and India.* 2 vols. Translated by E. Iliff Robson. Cambridge: Harvard University Press, 1949.

Azzi, Corry. "More on India's Sacred Cattle," *Current Anthropology* 15 (1974):317–324.

Babb, Lawrence A. "The Food of the Gods in Chhattisgarh: Some Structural Features of Hindu Ritual," *Southwestern Journal of Anthropology* 26 (1970):287–304.

Bajaj, Radha Krishna. "Making Maximum Use of Goshalas and Their Resources for the 4th Plan," *Gosamvardhana* 12, no. 5 (August 1964):22–24.

Balfour, Edward. *The Cyclopaedia of India and Southern Asia, Commercial, Industrial and Scientific: Products of the Mineral, Vegetable and Animal Kingdoms, Useful Arts and Manufactures.* 3 vols. 3d ed. London: Bernard Quaritch, 1885.

Banerjee, Narayanchandra. *Economic Life and Progress in Ancient India, Being the Outlines of an Economic History of Ancient India.* 2d ed. Calcutta: University of Calcutta Press, 1945.

Barthélemy Saint-Hilaire, J. *Hioeun-Thsang in India.* Translated from the French by Laura Enser. Calcutta: Susil Gupta, 1952.

Basham, A. L. *History and Doctrines of the Ājīvikas.* London: Luzac, 1951.

———. *The Wonder That Was India: A Survey of the Culture of the Indian Sub-Continent before the Coming of the Muslims.* London: Sidgewick and Jackson, 1954.

Baumgartel, Elise J. *The Cultures of Prehistoric Egypt.* 2 vols. London: Oxford University Press, 1960–1965.

Beal, Samuel (trans.). *Travels of Fah-Hian and Sung Yun.* London: Trübner, 1869.

———. *Si-Yu-Ki: Buddhist Records of the Western World.* London: Kegan Paul, Trench, Trübner and Co., 1884.

Bennett, John W. "On the Cultural Ecology of Indian Cattle," *Current Anthropology* 8 (1967):251–252.

———. "Comment on: An Approach to the Sacred Cow of India, by Alan Heston," *Current Anthropology* 12 (1971):197.

———. *The Ecological Transition: Cultural Anthropology and Human Adaptation.* New York: Pergamon Press, 1976.

Bernier, François. *Travels in the Moghul Empire.* Translated by A. Constable. Edited by V. A. Smith. London: Oxford University Press, 1934.

Bhaktivedanta Swami Prabhupada, A. C. *Kṛṣṇa: the Supreme Personality of Godhead. A Summary Study of Śrīla Vyāsadeva's Srīmad-Bhāgavatam, 10th Canto.* 3 vols. Boston, Massachusetts: Iskcon Press, 1970.

Bhandarkar, D. R. *Aśoka.* The Carmichael Lectures, 1923. Calcutta: University of Calcutta Press, 1955.

Bhandarkar, Ramakrishna G. *Vaishnavism, Shaivism and Minor Religious Systems.* Strassburg: K. J. Trübner, 1913.

Bhardwaj, Surinder Mohan. *Hindu Places of Pilgrimage in India.* Berkeley and Los Angeles: University of California Press, 1973.

Bhattacharya, P. "From Abu to Aarey," *Gosamvardhana* 11, nos. 7–8 (October–November 1963):6–12.

Bhattacharyya, Haridas (ed.). *The Cultural Heritage of India.* 4 vols. Rev. ed. Calcutta: Ramakrishna Mission, Institute of Culture, 1953–1961.

Bishop, C. W. "The Ritual Bullfight," *China Journal of Science and Arts* 3 (1925): 630–37.

Blunt, Sir Edward (ed.). *Social Service in India: An Introduction to Some Social and Economic Problems of the Indian People.* London: His Majesty's Stationery Office, 1939.

Bombay (Presidency). *Gazetteer of the Bombay Presidency. Vol. 4: Ahmedabad.* Bombay: Government Central Press, 1879.

———. *Gazetteer of the Bombay Presidency. Vol. 8: Kathiawar.* Bombay: Government Central Press, 1884.

———. *Gazetteer of the Bombay Presidency. Vol. 18: Poona, Part 3.* Bombay: Government Central Press, 1885.

———. *Gazetteer of the Bombay Presidency. Vol. 26: Bombay.* Bombay: Government Central Press, 1893–1894.

Bose, Atindra Nath. *Social and Rural Economy of Northern India 600* B.C.*–200* A.D. 2 vols. Calcutta: Firma K. L. Mukhopadhyay, 1961.

Boulais, G. *Manuel du Code Chinois.* Chang-Hai: Imprimerie de la Mission Catholique, 1923.

Brown, W. Norman. "The Sanctity of the Cow in Hinduism," *The Madras University Journal* 28 (January 1957):29–49; reprinted in *Economic Weekly* (Bombay) 16 (1964):245–255.

Buchanan, B. "A Dated Seal Impression Connecting Babylonia and Ancient India," *Archaeology* 20 (1967):104–107.

Bühler, Georg (trans.). *The Sacred Laws of the Âryas, as Taught in the Schools of Âpastamba, Guatama, Vâsishtha, and Baudhâyana.* Part 2: *Vâsishtha and Baudhayana.* Vol. 14 of *The Sacred Books of the East.* Edited by F. Max Müller. Oxford: The Clarendon Press, 1882.

——— (trans.). *The Laws of Manu.* Translated with extracts from seven commentaries. Vol. 25 of *The Sacred Books of the East.* Edited by F. Max Müller. Oxford: The Clarendon Press, 1886.

——— (trans.). *The Sacred Laws of the Âryas, as Taught in the Schools of Âpastamba, Gautama, Vâsishtha and Baudhâyana.* Part 1: *Âpastamba and Gautama.* Vol. 2 of *The Sacred Books of the East.* Edited by F. Max Müller. Oxford: The Clarendon Press, 1896.

———. *The Indian Sect of the Jainas.* Calcutta: Susil Gupta (India) Private Ltd., 1963.

Burnes, Alexander. "Art. VIII. Notice of a remarkable Hospital for Animals at

Surat. By Lieut. Alexander Burnes of the Bombay Military Establishment: being an Extract from a Manuscript Journal," *Journal of the Royal Anthropological Society of Great Britain and Northern Ireland* 1 (1834):96.

Chakravarti-Nayanar, Appasvami. *The Religion of Ahiṃsā: the Essence of Jaina Philosophy and Ethics.* Bombay: R. Hirachand, 1957.

Chatterjee, Satischandra. "Hindu Religious Thought," in *The Religions of the Hindus.* Edited by Kenneth W. Morgan. New York: The Ronald Press Co., 1953.

Chatterjee, S. "India's Cows and Plough Cattle and their Interrelation with Work and Milk Production," *Indian Agriculturalist* 7 (1963):13–22.

Chauhan, B. R. *A Rajasthan Village.* New Delhi: Vir Publishing House, 1967.

Chavda, Vijaysingh. *Bibliography of Gujarat.* Ahmedabad: New Order Book Co., 1972.

Chavarria-Aguilar, O. L. (ed.). *Traditional India.* Englewood Cliffs, New Jersey: Prentice-Hall, Inc., 1964.

Choudhuri, Narendra Nath. "Mother Goddess Durga," *The Poona Orientalist* 15 (1950):32–38.

Childe, V. Gordon. *The Aryans.* New York: Alfred A. Knopf, 1926.

Commissariat, M. S. *A History of Gujarat.* 3 vols. Bombay: Orient Longmans, 1938–1957.

Conrad, Jack Randolph. *The Horn and the Sword; The History of the Bull as Symbol of Power and Fertility.* New York: E. P. Dutton, 1957.

Conze, Edward. *Buddhist Meditations.* London: George Allen and Unwin Ltd., 1956.

———. *Buddhism: Its Essence and Development.* New York: Harper and Brothers, 1959.

———. *Buddhist Thought in India.* London: George Allen and Unwin Ltd., 1962.

Cover, John H., et al. *The Economy of India.* Subcontractor's Monograph, HRAF 32 California 1. 2 vols. Berkeley: Human Relations Area Files, South Asia Project, University of California, 1956.

Crooke, William. *The Popular Religion and Folk-Lore of Northern India.* 2 vols. Westminster: Archibald Constable and Co., 1896.

———. *The North-Western Provinces of India: Their History, Ethnology and Administration.* London: Methuen and Co., 1897.

———. *Things Indian, Being Discursive Notes on Various Subjects Connected with India.* London: John Murray, 1906.

———. "The Veneration of the Cow in India," *Folk-Lore* 23 (1912):275–306.

Cunningham, Sir Alexander. *The Ancient Geography of India.* Varanasi: Indological Book House, 1963.

Dales, George F., "Shifting Trade Patterns between the Iranian Plateau and the Indus Valley in the Third Millennium B.C.," *Colloques Internationaux du Cen-*

tre National de la Recherche Scientifique. No. 567. *Le Plateau Iranien et L'Asie Centrale des Origines à la Conquête Islamique,* 1976, pp. 67–78.

)almir, Jayadal. *A Review of 'Beef in Ancient India.'* Gorakhpur, Uttar Pradesh: Gita Press, 1971.

)andekar, V. M. "Problem of Numbers in Cattle," *Economic Weekly* 16 (1964): 331–353, 355.

——. "Cow Dung Models," *Economic and Political Weekly* 4 (1969):1267–1269, 1271.

——. "India's Sacred Cattle and Cultural Ecology," *Economic and Political Weekly* 4 (1969):1559–1567.

——. "Sacred Cattle and More Sacred Production Functions," *Economic and Political Weekly* 5 (1970):527, 529–531.

)ange, S. A. *Pastoral Symbolism from the Ṛgveda.* Poona: Poona University Press, 1970.

)arling, Malcolm Lyall. *Wisdom and Waste in the Punjab Village.* London: Humphrey Milford, Oxford University Press, 1934.

)armesteter, James (trans.). *The Zend-Avesta.* Pt. I. *The Vendidad.* Vol. 4 of *The Sacred Books of the East.* Edited by F. Max Müller. Oxford: Oxford University Press, 1880.

)as, S. K. "A Study of Folk Cattle-Rites," *Man in India* 33 (July–September 1953): 232–241.

)as Gupta, Satish C. *The Cow in India.* 2 vols. Calcutta: Khadi Pratisthan, 1945.

)as Gupta, Surendra Nath. *A History of Indian Philosophy.* 5 vols. Cambridge: Cambridge University Press, 1922–1955.

)avies, William Lewis. *Indian Indigenous Milk Products.* Calcutta: Thacker, Spink and Co., 1940.

)e Bary, William Theodore (ed.). *Sources of Indian Tradition.* 2 vols. New York: Columbia University Press, 1966.

)ebysingh, Molly. "Poultry and Cultural Distributions in India." Ph.D. dissertation, Syracuse University, 1970.

)evaraj, N. J. "*Ahiṃsā* in the Indian Tradition," *Indian Philosophy and Culture* 15 (March 1970):22–25.

)evi, Rukmini. "Cow-Slaughter," *Animal Citizen (Madras)* 4, no. 1 (October–November–December 1966):9–12.

)havamony, Mariasusai. *Love of God According to Śaiva Siddhanta: A Study in the Mysticism and Theology of Śaivism.* Oxford: The Clarendon Press, 1971.

)hebar, U. N. "The Cow," *Gosamvardhana* 9, nos. 7–8 (October–November 1961):5–7.

——. "The Central Council of Gosamvardhana: A Coordinated Effort at Cattle Development," *Gosamvardhana* 10, nos. 7–8 (October–November 1962):5–8.

Diener, Paul, Donald Nonini, and Eugene E. Robkin. "The Dialectics of the Sacre Cow: Ecological Adaptation vs. Political Appropriation in the Origins o India's Cattle Complex," *Dialectical Anthropology* 3 (1978):221–241.

Diener, Paul, and Eugene E. Robkin. "Ecology, Evolution, and the Search fo Cultural Origins: The Question of Islamic Pig Prohibition," *Current Anthro pology* 19 (1978):493–540.

Dikshitar, V. R. Ramachandra. "The Cow in Hindu Life," *The Journal of th Banaras Hindu University* 11, 3 (1938):271–282.

———. "A Note on Cow Veneration in Ancient India," pp. 75–77 in *A Volum of Eastern and Indian Studies*. Edited by S. M. Katre and P. K. Gode. Extr Series I of the *New Indian Antiquary*. Bombay: Karnatak Publishing House 1939.

Diwaker, S. C. *Glimpses of Jainism*. Delhi: Jain Mitra Mandal, Publicatio no. 163, 1964.

Dodwell, Henry H. (ed.). *The Cambridge History of India*. 5 vols. Cambridge Cambridge University Press, 1922–1953.

Dubois, Abbé J. A. *Hindu Manners, Customs and Ceremonies*. Translated an edited by Henry K. Beauchamp. 3d ed. Oxford: The Clarendon Press, 1906.

Dunbar, Gary S. "Ahimsa and Shikar: Conflicting Attitudes Towards Wildlif in India," *Landscape* 19, no. 1 (1970):24–27.

During-Caspers, E. C. L. "Further Evidence for Cultural Relations between India Baluchistan, Iran and Mesopotamia in Early Dynastic Times," *Journal of Nea Eastern Studies* 24 (1965):53–56.

Durrani, F. A. "Stone Vases as Evidence of Connection between Mesopotami and the Indus Valley," *Ancient Pakistan* 1 (1964):51–96.

Dutt, Romesh C. *A History of Civilisation in Ancient India Based on Sanskr Literature*. 2 vols. London: Kegan Paul, Trübner and Co., 1893.

Ellefsen, Richard A. "The Milk Supply of Major Indian Cities," Ph.D. disserta tion, University of California, Berkeley, 1968.

Elmore, Wilber Theodore. "Dravidian Gods in Modern Hinduism: A Study o Local and Village Deities of Southern India," *The University Studies of th University of Nebraska* 15 (1915):1–149.

Enthoven, R. E. *The Tribes and Castes of Bombay*. 3 vols. Bombay: Governmen Central Press, 1920–1922.

———. *The Folklore of Bombay*. Oxford: The Clarendon Press, 1924.

Fábri, C. L. "The Cretan Bull-Grappling Sports and the Bull Sacrifice in the Indu Valley Civilization," *Annual Report of the Archaeological Survey of India 1934–35*. 93–100.

Fārūqī, Isma'īl Rāgīal, and David E. Sopher (eds.). *Historical Atlas of the Rel gions of the World*. New York: Macmillan, 1974.

Forbes, Alexander Kinloch. *Rās Māla*. London: Oxford University Press, 1924.

Forbes, James. *Oriental Memoirs.* London: Richard Bentley, 1834.

Ford Foundation. *Report on India's Food Crisis and Steps to Meet It.* New Delhi: Ministry of Food and Agriculture and Ministry of Community Development and Cooperation, Government of India, 1959.

Foster, William (ed.). *Early Travels in India, 1583–1619.* London: Oxford University Press, 1921.

Freed, Stanley A., and Ruth S. Freed. "Cattle in a North Indian Village," *Ethnology* 11 (1972):399–408.

Fryer, John. *A New Account of East India and Persia, Being Nine Years' Travels 1672–1681.* 3 vols. Edited by William Crooke. London: The Hakluyt Society, Second Series, 1909–1915.

Gadd, C. J. "Seals of Ancient Indian Style found at Ur," *Proceedings of the British Academy* 17 (1932):191–210.

Gandhi, M. K. *How to Serve the Cow.* Ahmedabad: Navajivan Publishing House, 1954.

———. *India's Food Problem.* Ahmedabad: Navajivaṅ Publishing House, 1960.

Ganguli, R. "Cattle and Cattle Rearing in Ancient India," Bhandarkar Oriental Research Institute, Poona. *Annals* 12 (1931):216–230.

Geden, A. S. "Upanishads," pp. 540–548 in *Encyclopaedia of Religion and Ethics,* vol. 12. Edited by James Hastings. New York: Charles Scribner's Sons, 1955.

Gibb, H. A. R., and J. H. Kranes (eds.). *Shorter Encyclopedia of Islam.* Ithaca, New York: Cornell University Press, 1965.

Gillion, Kenneth L. *Ahmedabad. A Study in Indian Urban History.* Berkeley and Los Angeles: University of California Press, 1968.

Glacken, Clarence J. *Traces on the Rhodian Shore.* Berkeley and Los Angeles: University of California Press, 1967.

Glasenapp, Helmuth von. *The Doctrine of Karma in Jain Philosophy.* Bombay: Trustees, Bai Vijibai Jivanlal Panal Charity Fund, 1942.

———. *Der Jainismus, eine Indische Ertosangsreligion, nach den Quelledargestellt.* Hildesheim: Olms, 1964.

Gonda, J. *Aspects of Early Viṣṇuism.* Utrecht: A. Oosthoek, 1954.

———. *Viṣṇuism and Śivaism: A Comparison.* London: Athlone Press, 1970.

Goswami, C. L. (trans.). *Srīmad Bhāgavata Mahāpurāṇa.* Gorakhpur, Uttar Pradesh: Gita Press, 1971.

Gourou, Pierre. "Civilisation et Économie Pastorale," *L'Homme* 1963:123–129.

Growse, F. S. *Mathura—A District Memoir.* Allahabad: The Northwestern Provinces and Oudh Government Press, 1883.

Gubernatis, Angelo de. *Zoological Mythology, or the Legends of Animals.* 2 vols. London: Trübner & Co., 1872.

Guérinot, Armand Albert. *La Religion Djaina; Histoire, Doctrine, Cult, Coutumes, Institutions.* Paris: P. Geuthner, 1926.

Gujarat (State). Gazetteer of India. *Gujarat State Gazetteers.* Vol. 1: *Ahmedabad.* Ahmedabad: Director, Government Printing, Stationery and Publications, 1965.

———. *Gujarat State Gazetteers.* Vol. 14: *Rajkot.* Ahmedabad: Director, Government Printing, Stationery and Publications, 1965.

———. General Administration Department (Planning). *Fourth Five Year Plan, 1969–1974.* Ahmedabad, 1965.

———. State Archives, Ahmedabad. *Report on the Administration of Bhavnagar State for the Year 1911–1912.* Bhavnagar: Government Press, 1911.

———. *Report on the Administration of Morvi State for the Year 1934–35.* Morvi: Government Press, 1935.

———. Superintendent of Census Operations. *Census of India, 1961. District Census Handbooks.* Vol. 2: *Rajkot.* Ahmedabad: Director, Government Printing, Stationery and Publications, 1964.

———. *Census of India, 1961. District Census Handbooks.* Vol. 11: *Ahmedabad.* Ahmedabad: Director, Government Printing, Stationery and Publications, 1965.

Gupta, Shakti M. *Vishnu and His Incarnations.* Bombay and New Delhi: Somaiya Publications, 1974.

Hahn, Eduard. *Die Haustiere und ihre Beziehungen zur Wirtschaft des Menschen.* Leipzig: Duncker and Humblot, 1896.

Hall, H. R., and C. L. Woolley. *Ur Excavations.* Vol. I: *Al-Ubaid.* London: Oxford University Press, 1927.

Hamilton, Walter. *A Geographical, Statistical and Historical Description of Hindoostan and the Adjacent Countries.* 2 vols. London: John Murray, 1820.

Hanumantha Rao, C. H. "India's 'Surplus' Cattle: Some Empirical Results," *Economic and Political Weekly* 5 (1969):A225–227.

Haryana (State). Planning Department. *Fourth Five Year Plan 1969–74.* Chandigarh, 1971.

———. Planning Department. *Annual Plan Vol. 6, 1971–72.* Chandigarh, 1971.

———. Economics and Statistical Organization, Planning Department. *Basic Statistics on Plan Schemes.* Chandigarh, 1973.

Harper, Edward B. (ed.). *Religion in South Asia.* Seattle: University of Washington Press, 1964.

Harris, Marvin. "The Myth of the Sacred Cow," pp. 217–228 in *Man, Culture and Animals,* edited by Anthony Leeds and Andrew P. Vayda. Washington, D.C.: American Association for the Advancement of Science, 1965.

———. "The Cultural Ecology of India's Sacred Cattle," *Current Anthropology* 7 (1966):51–66.

———. *The Rise of Anthropological Theory: A History of Theories of Culture.* New York: Crowell, 1968.

———. *Cows, Pigs, Wars and Witches.* New York: Random House, 1974.

———. *Cannibals and Kings.* New York: Random House, 1977.

———. "India's Sacred Cow," *Human Nature* 1, no. 2 (1978):28–36.

———. *Cultural Materialism: The Struggle for a Science of Culture.* New York: Random House, 1979.

Hastings, James (ed.). *Encyclopaedia of Religion and Ethics.* 13 vols. New York: Charles Scribner's Sons, 1955.

Heston, Alan. "An Approach to the Sacred Cow of India," *Current Anthropology* 12 (1971):191–209.

Hiltebeitel, Alf. "The Indus Valley 'Proto-Śiva,' Re-examined through Reflections on the Goddess, the Buffalo, and the Symbol of *vāhanas,*" *Anthropos* 73 (1978): 767–797.

Himachal Pradesh (State). Animal Husbandry Department. *The Dying Cow: Can It Survive?* Simla: Animal Husbandry Department, 1963.

Hoffpauir, Robert. "India's Other Bovine: A Cultural Geography of the Water Buffalo," Ph.D. dissertation, University of Wisconsin, 1974.

———. "The Indian Milk Buffalo: A Paradox of High Performance and Low Reputation," *Asian Profile* 5 (1977):111–134.

———. "Subsistence Strategy and Its Ecological Consequences in the Nepal Himalaya," *Anthropos* 73 (1978):215–252.

Hopkins, E. Washburn. "The Social and Military Position of the Ruling Caste in Ancient India, as Represented by the Sanskrit Epic," *Journal of the American Oriental Society* 13 (1899):57–376.

———. *Epic Mythology.* Grundiss der Indo-Arischen Philologie und Altertumskunde. Band III, Heft I. B. Strassburg: Verlag Von Karl J. Trübner, 1919.

———. *Ethics of India.* New Haven: Yale University Press, 1924.

Hulsewe, A. F. P. *Periodieke executie- en Slacht-Verboden in de T'ang tijd en Hun oorsprong.* Leiden: E. J. Brill, 1948.

Hutton, J. H. *Caste in India: Its Nature, Function and Origins.* 3d ed. Bombay: Oxford University Press, 1961.

Ibn Batuta. *Ibn Batuta's Travels.* 3 vols. Translated by H. A. R. Gibb. London: The Hakluyt Society, 1929–1962.

India. Census Commissioner. *Census of India, 1891. General Tables for British Provinces and Feudatory States.* Vol. I. *Statistics of Area; Population; Towns and Villages; Religion; Age; Civil Condition; Literacy; Parent-Tongue; Birthplace; Infirmities and Occupation.* London, 1892.

———. *Census of India, 1901.* Vol. 5: *Assam.* Bombay: Government Central Press, 1902–1905.

India. *Imperial Gazetteer of India.* 26 vols. Oxford: The Clarendon Press, rev. ed., 1907–1909.

———. *Imperial Gazetteer of India. Provincial Series.* 25 vols. Calcutta: Superintendent of Government Printing, 1908–1909.

India (Republic). *Constitution of India.* New Delhi: Government of India Press, 1949.

India (Republic). Central Council of Gosamvardhana. *Report on the Survey of Sites for Gosadans.* New Delhi: Central Council of Gosamvardhana, 1956.

———. *Survey of Areas of Surplus Grass Production.* New Delhi: Central Council of Gosamvardhana, 1956.

———. *First Gosamvardhana Seminar (Srinagar).* 28 August–1 September 1958. New Delhi: Central Council of Gosamvardhana, 1961.

———. *Annual Report for the Year Ending 31st March, 1961.* New Delhi: Central Council of Gosamvardhana, 1961
1961.

———. *Second Gosamvardhana Seminar (Mount Abu).* 20–22 June 1960. New Delhi: Central Council of Gosamvardhana, 1962.

———. *Report of the Special Committee on Preserving High-yielding Cattle.* New Delhi: Central Council of Gosamvardhana, 1962.

———. *Annual Report for the Year Ending 31st March, 1962.* New Delhi: Central Council of Gosamvardhana, 1962.

———. *Annual Report for the Year Ending 31st March, 1963.* New Delhi: Central Council of Gosamvardhana, 1963.

———. *About the Central Council of Gosamvardhana.* New Delhi: Central Council of Gosamvardhana, 1963.

———. *Annual Report for the Year Ending 31st March, 1965.* New Delhi: Central Council of Gosamvardhana, 1965.

———. *Annual Report for the Year Ending 31st March, 1966.* New Delhi: Central Council of Gosamvardhana, 1966.

———. *Gosamvardhana Seminar on Approach to Cattle Development in the Country (Bombay).* New Delhi: Central Council of Gosamvardhana, 1966.

———. *Annual Report for the Year Ending 31st March, 1967.* New Delhi: Central Council of Gosamvardhana, 1967.

India (Republic). Indian Council of Agricultural Research. *1st Indian Dairy Yearbook, 1960.* New Delhi: Indian Council of Agricultural Research, 1961.

———. *Agriculture in Ancient India.* New Delhi: Indian Council of Agricultural Research, 1964.

India (Republic). Ministry of Food and Agriculture. *Report of the Expert Committee on the Prevention of Slaughter of Cattle in India.* Delhi: Manager of Publications, 1955.

India (Republic). Ministry of Food and Agriculture. Department of Agriculture. *Report on the Committee for the Prevention of Cruelty to Animals.* Delhi: Manager of Publications, 1957.

India (Republic). Ministry of Food and Agriculture. Directorate of Marketing and Inspection. *Agricultural Marketing in India. Report on the Marketing of Cattle in India.* Delhi: Manager of Publications, 1956.

India (Republic). Ministry of Food, Agriculture, Community Development and Cooperation. Directorate of Economics and Statistics. *Indian Livestock Census, 1961.* Delhi: Manager of Publications, 1969.

India (Republic). Ministry of Information and Broadcasting. *India: A Reference Annual, 1964.* Delhi: Director, Publications Division Ministry of Information and Broadcasting, 1964.

India (Republic). Ministry of Planning. Department of Statistics. Central Statistical Organization. *Statistical Abstract. India, 1972.* Delhi: Controller of Publications, 1974.

India (Republic). National Council of Applied Economic Research. *Domestic Fuel Consumption in Rural India.* New Delhi: National Council of Applied Economic Research, 1965.

India (Republic). Office of the Registrar General. *Census of India, 1961.* Vol. 1. *India.* Pt. 2-A(ii). *Union Primary Census Abstracts.* Delhi: Manager of Publications, 1963.

———. *Census of India, 1961.* Vol. 1: *India.* Pt. 2-A(i). General Population Tables. Delhi: Manager of Publications, 1964.

———. *Census of India, 1971.* Series 3: *Assam.* Pt. VI-A: *Town Directory.* Delhi: Controller of Publications, 1974.

———. *The First Five Year Plan.* New Delhi: Planning Commission, 1952. Planning Commission, 1952.

———. *Second Five Year Plan.* New Delhi: Planning Commission, 1956.

———. *Review of the First Five Year Plan.* New Delhi: Planning Commission, 1957.

———. *Third Five Year Plan.* Delhi: Manager of Publications, 1961.

———. *The Third Plan Mid-Term Appraisal.* Delhi: Manager of Publications, 1963.

———. *Annual Plan 1966–67.* Delhi: Manager of Publications, 1967.

India (Republic). Planning Commission. *Annual Plan 1967–68.* Delhi: Manager of Publications, 1968.

———. *Annual Plan 1968–69.* Delhi: Manager of Publications, 1969.

———. *Fourth Five Year Plan 1969–74.* Delhi: Manager of Publications, 1970.

———. *Draft Fifth Five Year Plan 1974–79.* Delhi: Manager of Publications, 1974.

Indian Famine Inquiry Commission. *The Famine Inquiry Commission, Fina Report.* Delhi: Manager of Publications, 1945.

Jacobi, Hermann. "Introduction" to *Jaina Sûtras.* Vol. 22 of *The Sacred Books c the East.* Edited by F. Max Müller. Oxford: The Clarendon Press, 1884.

——— (trans). *Jaina Sûtras, Part I and Part II.* Vols. 22 and 45 of *The Sacre Books of the East.* Edited by F. Max Müller. Oxford: The Clarendon Press 1884, 1895.

———. "Cow (Hindu)," pp. 224–226 in *Encyclopaedia of Religion and Ethics* vol. 4. Edited by James Hastings. New York: Charles Scribner's Sons, 1955.

———. "Durga," pp. 117–119 in *Encyclopaedia of Religion and Ethics,* vol. 5 Edited by James Hastings. New York: Charles Scribner's Sons, 1955.

Jain, Chote Lal. *Jaina Bibliography.* Calcutta: S. C. Seal, 1945.

Jain, Jagdish Chandra. *Life in Ancient India as Depicted in Jain Canons.* Bombay: New Book Co., 1947.

Jain, Jyoti Prasad. *The Jaina Sources of the History of Ancient India (100* B.C.–A.D. *900).* Delhi: Munshi Ram Manohar Lal, 1964.

Jain, Kailash Chand. *Jainism in Rajasthan.* Sholapur: Gulabchand Hirachan Doshi, 1963.

Jain, Jagmandar Lal. *Outlines of Jainism.* Cambridge: Cambridge Universit Press, 1916.

Jaiswal, Suvira. *The Origin and Development of Vaiṣṇavism.* New Delhi: Mirshiram Manoharlal, 1967.

James, E. O. *Myth and Ritual in the Ancient Near East; An Archaeological an Documentary Study.* New York: Frederick A. Praeger, 1958.

———. *The Cult of the Mother-Goddess; An Archaeological and Documentar Study.* New York: Frederick A. Praeger, 1959.

———. *The Ancient Gods; The History and Diffusion of Religion in the Ancien Near East and Eastern Mediterranean.* New York: G. P. Putnam's Sons, 1960.

———. *The Worship of the Sky-God; A Comparative Study in Semitic and Indo European Religion.* London: The Athlone Press, 1963.

Jangam, R. T., and G. Thimmaiah. *Economics and Politics of Cow-Slaughter Ban* Mysore: Sharat Prakashana, 1967.

Jani, Dahyabhai H. *Romance of the Cow.* Bombay: Bombay Humanitaria League, 1938.

Jhaveri, Krishnalal Mohanlal (trans.). *Imperial Farmans (*A.D. *1577 to* A.D. *1805 Granted to the Ancestors of His Holiness the Tikayat Maharaj.* Translated int English, Hindi, and Gujarati by Krishnalal Mohanlal Jhaveri. Bombay: Th New Printing Press, n.d.

Kane, Pandurang Vaman. *History of Dharmaśāstra.* 5 vols. in 7 tomes. Poona Bhandarkar Oriental Research Institute, 1930–1962.

Kapadia, H. R. "Prohibition of Flesh-Eating in Jainism," *Review of Philosophy and Religion (Poona, India)* 4 (1933):232–239.

Kapp, K. William. *Hindu Culture, Economic Development and Economic Planning in India: A Collection of Essays.* Bombay: Asia Publishing House, 1963.

Kaura, R. L. *Cattle Development in Uttar Pradesh.* Allahabad: Department of Animal Husbandry, Uttar Pradesh, 1950.

———. "Gosadan: A House for Unwanted and Infirm Cattle," *Indian Livestock* 2, no. 3 (1964):11–12.

Keith, A. B. *Religion and Philosophy of the Vedas.* Cambridge, Mass.: Harvard University Press, 1925.

Khurody, D. N. *Dairying in India. A Review.* Bombay: Asia Publishing Co., 1974.

Kielhorn, F. "Three Inscriptions from Travancore," *Epigraphia Indica* 4, no. 27 (1896–1897):201–204.

Kinsley, David R. *The Sword and the Flute: Kālī and Kṛṣṇa, Dark Visions of the Terrible and the Sublime in Hindu Mythology.* Berkeley and Los Angeles: University of California Press, 1975.

Kosambi, Damodar Dharmanand. *Myth and Reality: Studies in the Formation of Indian Culture.* Bombay: Popular Prakashan, 1962.

Koskenniemi, A., A. Parpola, and S. Parpola. "Materials for the Study of the Indus Script I: A Concordance of the Indus Inscriptions," *Annales,* Academiae Scientariarum Fennicae, Series B, vol. 185.

Kothavala, Zal R. "Milk Production in India," *Agriculture and Livestock in India* 2 (1932):122–129.

Kotturan, George, *Ahiṃsā: Gautama to Gandhi.* New Delhi: Sterling Publishers, 1973.

Kramer, Samuel Noah. *Sumerian Mythology.* New York: Harper and Row, 1961.

———. *The Sumerians.* Chicago: University of Chicago Press, 1963.

Kulke, Eckehard. *The Parsees in India: A Minority as Agent of Social Change.* Munchen: Weltforum Verlag, 1974.

Lakshminarayan Rao, N. "Two Stone Inscriptions of Krishna II; Saka 805," *Epigraphia Indica* 21, no. 35 (1931–32):205–209.

Lal, Mukandi. "Cow Cult in India," pp. 15–34 in *Cow-Slaughter: Horns of a Dilemma.* Edited by A. B. Shah. Bombay: Lalvani Publishing House, 1967.

Lall, H. K. "Myth of Surplus Cattle, Pt. I," *Gosamvardhana* 11, no. 6 (September 1963):25–29.

———. "Myth of Surplus Cattle, Pt. II," *Gosamvardhana* 11, no. 9 (December 1963):9–15.

———. *The Resurrection of the Cow in India.* Hoshiarpur: Vishveshnaranand Institute, 1973.

Lamberg-Karlovsky, C. C. "Trade Mechanisms in Indus-Mesopotamian Interrelations," *Journal of the American Oriental Society* 92, no. 2 (1972):222–229.

Langlois, M. A. (trans.). *Harivanśa ou Histoire de la Famille de Hari, ouvrage formant un appendice du Mahābhārata.* 2 vols. Paris: Oriental Translation Fund of Great Britain and Ireland, 1835.

Laufer, Berthold. "Methods in the Study of Domestications," *Scientific Monthly* 25 (1927):251–255.

Leeds, Anthony, and Andrew P. Vayda (eds.). *Man, Culture and Animals: The Role of Animals in Human Ecological Adjustments.* Washington, D.C.: American Association for the Advancement of Science, 1965.

Legge, James (trans.). *A Record of Buddhistic Kingdoms, Being an Account by the Chinese Monk Fa-Hien of His Travels in India and Ceylon* (A.D. *399–414*) *in Search of the Buddhist Books of Discipline.* New York: Paragon Book Reprint Corp. and Dover Publications, reprint of the 1886 edition, 1965.

Levy, E. R. *The Gate of Horn.* London: Faber and Faber, 1948.

Lewis, Oscar. *Village Life in Northern India: Studies in a Delhi Village.* Urbana Illinois: University of Illinois Press, 1958.

Linschoten, Jan Huyghen van. *The Voyage of John Huyghen van Linschoten to the East Indies.* 2 vols. Edited by A. C. Burnell and P. A. Tiele. London: The Hakluyt Society, 1885.

Littlewood, R. W. *Livestock of Southern India.* Madras: Superintendent, Government Press, 1936.

Locke, J. Courtenay (ed.). *The First Englishmen in India.* London: George Routledge and Sons, 1930.

Lodrick, Deryck O. "On Religion and Milk Bovines in an Urban Indian Setting," *Current Anthropology* 20 (1979):241–242.

Lorenzen, David N. *The Kāpālikas and Kālāmukhas: Two Lost Śaivite Sects* Berkeley and Los Angeles: University of California Press, 1972.

McCrindle, John W. *Ancient India as Described by Megasthenes and Arrian.* London. Trübner, 1877.

Macdonell, Arthur A. *Vedic Mythology.* Vol. 3, pt. 1A of the *Encyclopedia of Indo-Aryan Research.* Edited by G. Bühler. Varanasi: Rameshwar Singh 1896.

———, and A. B. Keith. *Vedic Index of Names and Subjects.* 2 vols. London John Murray, 1912.

Mackay, Ernest. *Further Excavations at Mohenjo-Daro.* Delhi: Manager of Publications, 1937–1938.

———. *Chanhu-Daro Excavations 1935–1936.* American Oriental Series, vol. 20 New Haven, Connecticut: American Oriental Society, 1943.

Maharashtra State Record Office, Bombay (unpublished records). *Bombay Government Files. Petitions,* 9 July 1827. General Department 13/146/1827.

———. *Petitions,* 8th October 1827. General Department 170/1828.

———. *Report on Town Duties* by J. Langford. Revenue Department 106/1069/1839.

Majumdar, Asoke Kumar. *Chaulukyas of Gujarat: A Survey of the History and Culture of Gujarat from the Middle of the Tenth to the End of the Thirteenth Century*. Bombay: Bharatiya Vidya Bhavan, 1956.

Majumdar, Bimanbehari. *Kṛṣṇa in History and Legend*. Calcutta: University of Calcutta, 1969.

Majumdar, D. N. "Acculturation among the Hajong of Meghalaya," *Man in India* 52, no. 1 (March 1972):46–63.

Majumdar, M. R. *Cultural History of Gujarat (From Early Times to Pre-British Period)*. Bombay: Popular Prakashan, 1965.

Majumdar, R. C. *The Classical Accounts of India*. Calcutta: Firma K. L. Mukhopadhyay, 1960.

———, and A. D. Pusalker (eds.). *The Vedic Age*. Vol. 1 of *The History and Culture of the Indian People*. London: George Allen and Unwin, 1951.

Makhijani, H. J. "Development and Utilization of Cattle in India," *Gosamvardhana* 9, nos. 7–8 (October–November 1961):7–8, 33–36.

———. *Gaushalas and Pinjrapoles*. New Delhi: Central Council of Gosamvardhana, 1963.

Malabari, Behramji M. *Gujarat and Gujaratis: Pictures of Men and Manners taken from Life*. Bombay: Education Society Press, 1884.

Mandelbaum, David G. *Society in India*. 2 vols. Berkeley and Los Angeles: University of California Press, 1970.

Mankar, J. N. "Improving Potentialities of *Goshalas* for Cattle Development Work in the 4th Five Year Plan," *Gosamvardhana* 12, no. 5 (August 1964):17–21.

Mann, Harold H. "The Supply of Milk to Indian Cities," *Agricultural Journal of India (Calcutta)* 9, pt. 2:160–177.

Manucci, Niccolo. *Storio do Mogor: or Mogul India*. 4 vols. Translated by W. Irvine. London: John Murray, 1906–1908.

Marriott, McKim (ed.). *Village India: Studies in the Little Community*. Chicago: University of Chicago Press, 1955.

Marshall, Sir John. *Mohenjo-Daro and the Indus Civilisation: Being an Official Account of Archaeological Excavations at Mohenjo-Daro Carried Out by the Government of India Between the Years 1922 and 1927*. 3 vols. London: Arthur Probsthain, 1931.

Mathur, K. S. *Caste and Ritual in a Malwa Village*. Calcutta: Asia Publishing House, 1964.

Mathur, P. D., and R. N. Mukerji. "Goshalas in Rajasthan," *Gosamvardhana* 11, no. 2 (May 1963):3–4.

Mayadas, C. *Between Us and Hunger*. London: Oxford University Press, 1954.

Mehta, Mohan Lal. *Jaina Culture*. Varanasi: P. V. Research Institute, 1969.

Mellaart, James. *Çatal Hüyük*. London: Thames and Hudson, 1967.

Mishra, K. C. *The Cult of Jagannatha*. Calcutta: Firma K. L. Mukhopadhyay, 1971

Mishra, Vikas. *Hinduism and Economic Growth*. London: Oxford Universit Press, 1962.

Mitra, Kalipada. "The Gāydāṇṛ Festival and Its Parallels," *Indian Antiquary* 6((October 1931):187–190 and 61 (December 1931):235–238.

Modi, J. J. "Ancient Pataliputra," *Journal of the Bombay Branch of the Roya Asiatic Society* 24 (1915–1917):457–532.

Morris, M. D. "Values as an Obstacle to Economic Growth in South Asia: A Historical Survey," *Journal of Economic History* 27 (1967):588–607.

Mukherji, Shyam Chand. *A Study of Vaiṣṇavism in Ancient and Mediaeval Benga up to the Advent of Caitanya (Based on Archaeological and Literary Data)* Calcutta: Punthi Pustak, 1966.

Mulji, Karsandas. *History of the Sect of Maharajas, or Vallabhacharyas ir Western India*. London: Trübner, 1865.

Mundy, Peter. *The Travels of Peter Mundy in Europe and Asia 1608–1667*. 5 vols Edited by Lt. Col. Sir Richard Carnac Temple, Bt., C.I.E. London: The Haklu Society, Second Series, 1907–1936.

Nair, Kusum. *Blossoms in the Dust: The Human Factor in Indian Development* New York: Frederick A. Praeger, 1962.

Narayan, Śriman. "Livestock Development in the Third Plan," *Indian Livestock* 1 no. 1 (January 1963):4–6.

Nikam, N. A., and Richard McKeon (eds.). *The Edicts of Asoka*. Chicago: Th University of Chicago Press, 1959.

Nilakanta Sastri, K. A. (ed.). *Age of the Nandas and Mauryas*. Varanasi: Motila Banarsidass, 1952.

Oaten, Edward Farley. *European Travellers in India*. London: Kegan Paul, Trench Trübner and Co., 1909.

Odend'hal, Stewart. "Gross Energetic Efficiency of Indian Cattle in Their Environ ment," *Journal of Human Ecology* 1 (1972):1–27.

O'Flaherty, Wendy Doniger. *Asceticism and Eroticism in the Mythology of Śiva* London: Oxford University Press, 1973.

Olver, Arthur. "The Inadequacy of the Dual-Purpose Animal as the Goal in Cattl Breeding in India," *Agriculture and Livestock in India* 6 (1936):389–396.

O'Malley, L. S. S. *Indian Caste Customs*. Cambridge: Cambridge University Press 1932.

Ovington, John. *A Voyage to Suratt in the Year 1689*. Edited by H. G. Rawlinson London: Oxford University Press, 1929.

Palmieri, Richard P. "The Yak in Tibet and Adjoining Areas: Its Economic Uses

Social Inter-Relationships, and Religious Functions." Master's thesis, University of Texas, Austin, 1970.

———. "The Domestication, Exploitation, and Social Functions of the Yak in Tibet and Adjoining Areas," *Proceedings of the Association of American Geographers* 4 (1972):80–83.

———. "Domestication and Exploitation of Livestock in the Nepal Himalaya and Tibet: An Ecological, Functional, and Culture Historical Study of Yak and Yak Hybrids in Society, Economy, and Culture." Ph.D. dissertation. University of California, Davis, 1976.

arnerkar, Y. M. *On Cattle Keeping*. New Delhi: Central Council of Gosamvardhana, n.d.

———. "New Vistas of Activity for Goshalas. Certified Milk Production," Editorial, *Gosamvardhana* 11, no. 11 (February 1964):2.

———. "New Vistas of Activities for Goshalas," Editorial, *Gosamvardhana* 11, no. 12 (March 1964):2.

———. "On Better Utilization of Resources of Goshalas and Other Institutions," *Gosamvardhana* 12, no. 5 (August 1964):25–27.

———. "On Goshalas," Editorial, *Gosamvardhana* 12, no. 5 (August 1964):2.

'arpola, Asko. "The Indus Script Decipherment. The Situation at the end of 1969," *Journal of Tamil Studies* 2, no. 1 (April 1970).

———, S. Koskenniemi, S. Parpola, and P. Aalto. *Further Progress in the Indus Script Decipherment*. Copenhagen: The Scandinavian Institute of Asian Studies, Special Publications no. 3, 1970.

'arrack, Dwain W. "An Approach to the Bioenergetics of Rural West Bengal," pp. 29–46 in *Environment and Cultural Behaviour*. Edited by Andrew P. Vayda. Garden City, New York: Natural History Press, 1969.

'aton, Lewis Bayles. "Ishtar," pp. 428–434 in *Encyclopaedia of Religion and Ethics*, vol. 7. Edited by James Hastings. New York: Charles Scribner's Sons, 1955.

'ocock, D. F. *Mind, Body and Wealth: A Study of Belief and Practice in an Indian Village*. Totowa, New Jersey: Rowman and Littlefield, 1973.

'olo, Marco. *The Book of Ser Marco Polo, the Venetian, Concerning the Kingdoms and Marvels of the East*. Translated and edited by Col. Sir Henry Yule. 3d ed., revised by Henri Cordier. 2 vols. London: John Murray, 1903.

'rakash, Om. *Food and Drinks in Ancient India (From Earliest Times to c. 1200* A.D.). Delhi: Munshi Ram Manohar Lal, 1961.

'urchas, Samuel. *Hakluytus Posthumus; or, Purchas His Pilgrimes, Contayning a History of the World in Sea Voyages and Lande Travells by Englishmen and Others*. 20 vols. Reprint of the 1625 ed. New York: Macmillan, 1905–1907.

ładhakrishnan, S. *Indian Philosophy*. 2 vols. New York: The Macmillan Company, 1923.

Raj, K. N. "Investment in Livestock in Agrarian Economies: An Analysis of Some Issues Concerning 'Sacred Cows' and 'Surplus Cattle,'" *Indian Economic Review* 4, series 2:53–85.

———. "India's Sacred Cattle. Theories and Empirical Findings," *Economic and Political Weekly* 6 (1971):717–722.

Rajputana (Province). *The Rajputana Gazetteer.* 3 vols. Simla: Government Central Branch Press, 1880.

Ramanujam, B. V. *History of Vaishnavism in South India up to Ramanuja.* Annamalainagar, Tamil Nadu: Annamalai University, 1973.

Rampuria, Shree Chand. *The Cult of Ahiṃsā: A Jain View-Point.* Calcutta: Sri Jain Swetamber Terapanthi Mahasabha, 1947.

Randhawa, M. S. "Role of Domesticated Animals in Indian History," *Science and Culture (Calcutta)* 12, no. 1 (1946):5–14.

Rao, S. R. *Lothal and the Indus Civilization.* New York: Asia Publishing House, 1974.

Ravenholt, Albert. *India's Bovine Burden.* American Universities Field Staff Reports Service, South Asia series 10, no. 12 (India). New York: American Universities Field Staff, Inc., 1966.

Raychaudhuri, H. C. *Materials for the Study of the Early History of the Vaishnava Sect.* Calcutta: University of Calcutta, 1936.

Reed, Charles A. "The Pattern of Animal Domestication in the Prehistoric Near East," pp. 361–380 in *The Domestication and Exploitation of Plants and Animals.* Edited by Peter J. Ucko and G. W. Dimbleby. Chicago: Aldine Publishing Co., 1969.

Renou, Louis. *The Civilization of Ancient India.* Translated by Philip Spratt. Calcutta: Susil Gupta (Private) Ltd., 1954.

Rhys Davids, T. W. (trans.). *Buddhist Suttas.* Vol. 11 of *The Sacred Books of the East.* Edited by F. Max Müller. Oxford: The Clarendon Press, 1881.

Riencourt, Amaury de. *The Soul of India.* New York: Harper and Brothers, 1960.

Risley, H. H., and E. A. Gait. *Census of India, 1901.* Vol. 1: *India.* Pt. 1: *Report* Calcutta: Office of the Superintendent of Government Printing, 1903.

Rousselet, Louis. *L'Inde des Rajahs.* Paris: Libraire Hachette et Cie., 1875.

Roy, Prodipto. "The Sacred Cow in India," *Rural Sociology* 20 (1955):8–15.

———. "Social Background," *Seminar (New Delhi)* no. 93 (May 1967). (Issue devoted to "The Cow: A Symposium on the Many Implications of a Current Agitation.")

Roy, Sarat Chandra. *Oraon Religion and Customs.* Calcutta: K. M. Banerjee, at the Industry Press, 1928.

Royal Commission on Agriculture in India. *Report of the Royal Commission on Agriculture in India.* Vol. 1A: *Abridged Report and Report.* London: His Majesty's Stationery Office, 1928.

Sahlins, Marshall. *Culture and Practical Reason.* Chicago: University of Chicago Press, 1976.

Sangave, Vilas. *A Jaina Community: A Social Survey*. Bombay: Popular Book Depot, 1959.

Sankalia, H. D. "(The Cow) In History," *Seminar (New Delhi)* no. 93 (May 1967): 12–16.

Sathe, S. P. "Cow-Slaughter: The Legal Aspect," pp. 69–82 in *Cow-Slaughter: Horns of a Dilemma.* Edited by A. B. Shah. Bombay: Lalvani Publishing House, 1967.

Sayce, A. H. "Bull (Semitic)," pp. 887–889 in *Encyclopaedia of Religion and Ethics,* vol. 2. Edited by James Hastings. New York: Charles Scribner's Sons, 1955.

Schneider, Burch H. "The Doctrine of *Ahiṃsā* and Cattle Breeding in India," *Scientific Monthly* 67 (1948):87–92.

Schubring, W. *The Doctrine of the Jainas.* Translated from the German edition by Wolfgang Beurlen. Varanasi: Motilal Banarsidass, 1962.

Schwabe, Calvin W. "Holy Cow—Provider or Parasite? A Problem for Humanists," *Southern Humanities Review* 13, no. 3 (1978):251–278.

Schweitzer, Albert. *Indian Thought and Its Development.* London: Adam and Charles Black, 1956.

Sen, A. *Elements of Jainism.* Calcutta: Indian Publicity Society, 1953.

Service, Elman R. "The Prime-Mover of Cultural Evolution," *Southwestern Journal of Anthropology* 24 (1968):397–409.

Shah, A. B. (ed.). *Cow-Slaughter: Horns of a Dilemma.* Bombay: Lalvani: Publishing House, 1967.

Shah, M. M. "Cow-Slaughter: The Economic Aspect," pp. 44–68 in *Cow-Slaughter: Horns of a Dilemma.* Edited by A. B. Shah, Bombay: Lalvani Publishing House, 1967.

Shamasastry, R. *Kauṭilya's Arthaśāstra.* 6th ed. Mysore: Mysore Printing and Publishing House, 1960.

Sharma, D. S. *The Gandhi Sūtras.* New York: Devin Adair, 1949.

Sharma, R. C. "Animal Figurines in Mathura Art," *Silver Jubilee Souvenir,* Uttar Pradesh College of Veterinary Science and Animal Husbandry, Mathura, n.d.

Sharma, S. R. *Jainism and Karnataka Culture.* Dharwar: M. S. Kamalpur, 1940.

Sheth, Chimanlal Bhailal. *Jainism in Gujarat.* Bombay: Shree Vijayadevsur Sangh Gnan Samiti, 1953.

Shrivasta, D. D. "Indian Dairy-Farming: Its Historical Background and Economic Trends," *Rural India* 17, no. 11 (December 1954):442–446.

Shukla, K. K. "Goshala Development Scheme. Gujarat State," *Gosamvardhana* 11, no. 1 (April 1963):31–33.

Simoons, Frederick J. *Eat Not This Flesh.* Madison, Wisconsin: University of Wisconsin Press, 1961.

———. *A Ceremonial Ox of India: The Mithan in Nature, Culture and History, with Notes on the Domestication of Common Cattle.* Madison, Wisconsin: University of Wisconsin Press, 1968.

———. "The Traditional Limits of Milking and Milk Use in Southern Asia," *Anthropos* 65 (1970):547–593.

———. "The Determinants of Dairying and Milk Use in the Old World: Ecological, Physiological and Cultural," *Ecology of Food and Nutrition* 2 (1973): 83–90.

———. "The Sacred Cow and the Constitution of India," *Ecology of Food and Nutrition* 2 (1973):281–295.

———. "The Purificatory Role of the Five Products of the Cow in Hinduism," *Ecology of Food and Nutrition* 3 (1974):21–34.

———. "Fish as Forbidden Food: The Case of India," *Ecology of Food and Nutrition* 3 (1974):185–201.

———. "Contemporary Research Themes in the Cultural Geography of Domesticated Animals," *Geographical Review* 64 (1974):557–576.

———. "Questions in the Sacred-Cow Controversy," *Current Anthropology* 20 (1979):467–493.

Singer, Milton, "Cultural Values in India's Economic Development," *Annals of the American Academy of Political and Social Science* 305 (1956):81–91.

——— (ed.). *Krishna: Myths, Rites and Attitudes.* Chicago: University of Chicago Press, 1966.

Singh, Harbans. "Role of Goshalas and Pinjrapoles in the Cattle Economy of India," *Indian Farming* 10 (1949):481–484.

———. *Gaushalas and Pinjrapoles in India.* New Delhi: Central Council of Gosamvardhana, 1955.

———. *A Handbook of Animal Husbandry for Extension Workers.* New Delhi: Ministry of Food and Agriculture, Directorate of Extension, 1963.

———, and Y. M. Parnerkar. *Basic Facts about Cattle Wealth and Allied Matters.* New Delhi: Central Council of Gosamvardhana, 1966.

Singh, Sardar Datar. *Reorganisation of Gaushalas and Pinjrapoles in India.* New Delhi: Ministry of Agriculture, 1948.

———. "Resources of Goshalas and Their Utilization for Cattle Development Work in the Fourth Plan," *Gosamvardhana* 12, no. 5 (August 1964):9–16.

Sinh Jee, H. H. Sir Bhagvat. *A Short History of Aryan Medical Science.* London: Macmillan and Co., 1896.

Sinha, Bashishtha Narayan. "Development of *Ahiṃsā* in the Vedic Tradition," *Prajna* 13, no. 2 (March 1968):145–158.

Sircar, Dines Chandra. "Early History of Vaiṣnavism," pp. 108–145 in *The Cultural Heritage of India,* vol. 4. Edited by Haridas Bhattacharyya. Calcutta: Ramakrishna Mission, Institute of Culture, 1956.

———. *Inscriptions of Aśoka.* Delhi: The Publications Division, Ministry of Information and Broadcasting, Government of India, 1957.

Sital Prasad, Brahmachari. *A Comparative Study of Jainism and Buddhism.* Madras: Jain Mission Society, 1934.

Skelton, Robert. *Rajasthani Temple Hangings of the Krishna Cult.* New York: The American Federation of Arts, 1973.

Smith, V. A. *Asoka.* Oxford: The Clarendon Press, 1920.

———. *The Early History of India, From 600 B.C. to the Muhammadan Conquest Including the Invasion of Alexander the Great.* Oxford: The Clarendon Press, 1924.

Sopher, David. "Pilgrim Circulation in Gujarat," *Geographical Review* 58 (1968): 392–425.

———. "Indian Pastoral Castes and Livestock Ecologies: A Geographic Analysis," pp. 183–208 in *Pastoralists and Nomads in South Asia.* Schriftenreihe des Südasien-Instituts der Universität Heidelberg. Edited by Lawrence Saadia Leshnik and Günther-Dietz Sontheimer. Wiesbaden: Otto Harrassowitz, 1975. witz, 1975.

Spate, O. H. K., and A. T. A. Learmonth. *India and Pakistan: A General and Regional Geography.* London: Methuen and Co., 1972.

Srinivas, M. N. *Caste in Modern India and Other Essays.* Bombay: Asia Publishing House, 1962.

Srinivasan, Doris. "The So-called Proto-Śiva Seal from Mohenjo-Daro: An Iconological Assessment," *Archives of Asian Art* 29 (1975–76):47–58.

Stevenson, Mrs. Sinclair. *The Heart of Jainism.* London: Oxford University Press, 1915.

———. "Svetambaras," pp. 123–124 in *Encyclopaedia of Religion and Ethics,* vol. 12. Edited by James Hastings. New York: Charles Scribner's Sons, 1955.

Steward, Julian H. *Theory of Culture Change: The Methodology of Multilinear Evolution.* Urbana, Illinois: University of Illinois Press, 1955.

Strommenger, Eva. *Five Thousand Years of the Art of Mesopotamia.* Translated by Christina Haglund. New York: Harry N. Abrams, 1964.

Subrahmanya Aiyer, K. V. "Tinnevelly Inscriptions of Maravarman Sundara-Pandya II," *Epigraphia Indica* 24, no. 22 (1937–38): 153–172.

Sullivan, H. P. "A Re-Examination of the Religion of the Indus Civilization," *History of Religions* 4 (1964):115–125.

Sundara Ram, L. L. *Cow Protection in India.* George Town, Madras: The South Indian Humanitarian League, 1927.

Tähtinen, Unto. *Ahiṃsā: Non-Violence in Indian Tradition.* London: Rider, 1976.

Tatia, Nathmal. *Studies in Jaina Philosophy.* Banaras: Jain Cultural Society, 1951.

Tavernier, Jean Baptiste. *Travels in India.* 2 vols. Translated by V. Ball. Edited by William Crooke. London: Oxford University Press, 1925.

Thakur, Upendra. *Studies in Jainism and Buddhism in Mithila.* Varanasi: Chowkhamba Sanskrit Series Office, 1964.

Thévenot, Jean de. *The Travels of Monsieur de Thévenot into the Levant.* Translated by A. Lovell. London: H. Clark, 1971.

Thomas, P. *Hindu Religious Customs and Manners.* Bombay: D. B. Taraporevala & Sons, n.d.

———. *Festivals and Holidays of India.* Bombay: D. B. Taraporevala & Sons, 1971.

Toothi, N. A. *The Vaishnavas of Gujarat.* London: Longmans, 1935.

Thureau-Dangin, F., and M. Dunand. *Til Barsib.* Paris: Libraire Orientaliste Paul Geuthner, 1936.

Tiemann, Gunter. "Cattle Herds and Ancestral Land Among the Jat of Haryana in Northern India," *Anthropos* 65 (1970):481–504.

Tiwari, R. D. *Indian Agriculture.* Bombay: New Book Company, 1943.

Tod, Lt. Col. James. *Annals and Antiquities of Rajsthan.* 3 vols. London: Oxford University Press, 1920.

Uttar Pradesh (State). *Gazetteer of India. Uttar Pradesh District Gazetteers.* Vol. 54: *Varanasi.* Lucknow: Superintendent of Printing and Stationery. 1965.

———. *Uttar Pradesh District Gazetteers.* Vol. 36: *Mathura.* Lucknow: Superintendent of Printing and Stationery, 1968.

Uttar Pradesh (State). Superintendent of Census Operations. *Census of India, 1961. District Census Handbooks.* Vol. 21: *Mathura.* Lucknow: Superintendent of Printing and Stationery, 1965–1966.

———. *Census of India, 1961. District Census Handbooks.* Vol. 53: *Varanasi.* Lucknow: Superintendent of Printing and Stationery, 1965–1966.

Valle, Pietro della. *The Travels of Pietro della Valle in India.* 2 vols. Edited by E. Grey. London: The Haklyut Society, 1892.

Vayda, Andrew P., and Roy A. Rappaport. "Ecology, Cultural and Noncultural," pp. 476–497 in *Introduction to Anthropology.* Edited by J. A. Clifton. Boston: Houghton Miflin, 1968.

Walker, Benjamin. *The Hindu World: An Encyclopaedic Survey of Hinduism.* 2 vols. New York: Frederick A. Praeger, 1968.

Walli, Kosheyla. *Ahiṃsā in Indian Thought.* Varanasi: Bharata Manisha, 1974.

Watters, Thomas (trans.). *On Yüan Chwang's Travels in India. 629–649* A.D. 2 vols. London: Royal Asiatic Society, 1904–1905.

Weber, Max. *The Protestant Ethic and the Spirit of Capitalism.* New York: Charles Scribner's Sons, 1958.

———. *The Religion of India: The Sociology of Hinduism and Buddhism.* Translated and edited by Hans H. Gerth and Don Martindale. New York: The Free Press, 1958.

Weizman, Howard. "More on India's Sacred Cattle," *Current Anthropology* 15 (1974):321–323.

Wheatley, Paul. "A Note on the Extension of Milking Practices into Southeast Asia during the First Millennium A.D.," *Anthropos* 60 (1965):577–590.

Whitehead, Henry. *The Village Gods of South India.* London: Oxford University Press, 1921.

Wilberforce-Bell, Capt. H. *The History of Kathiawad from the Earliest Times.* London: William Heinemann, 1916.

Wilson, Thomas. *The Swastika, the Earliest Known Symbol, and Its Migrations; with Observations on the Migration of Certain Industries in Prehistoric Times.* Washington, D.C.: Government Printing Office, 1896.

Winternitz, M. *A General Index to the Names and Subject Matter of the Sacred Books of the East.* Delhi: Motilal Banarsidass, 1966.

———. *A History of Indian Literature.* 2 vols. New York: Russell and Russell, 1971.

Woolley, Sir Charles Leonard. *Ur: The First Phases.* New York: Penguin Books, 1946.

———. *Excavations at Ur.* London: Barnes and Noble, 1954.

Xavier, P. A. "The Role of the Pinjrapoles in the Animal Welfare Work in India," *Animal Citizen (Madras)* 4, no. 3 (April–May–June 1967):35–38.

Yule, Henry, and A. C. Burnell. *Hobson-Jobson: A Glossary of Colloquial Anglo-Indian Words and Phrases, and of Kindred Terms, Etymological, Historical, Geographical and Discursive.* New edition edited by William Crooke. London: Routledge and Kegan Paul, 1903.

Zaehner, R. C. (trans.). *The Bhagavad-Gita.* Oxford: The Clarendon Press, 1969.

Zimmer, Heinrich. *Myths and Symbols in Indian Art and Civilization.* Edited by Joseph Campbell. Princeton, New Jersey: Princeton University Press, 1946.

INDIAN WORKS CITED

Agni Purāṇa

Aitareya Brāhmaṇa

Arthaśāstra

Atharva Veda

Baudhāyana Dharma Sūtra

Bhagavad Gītā

Bhāgavata Purāṇa

Bhagavatī Sūtra

Bhavisya Purāṇa

Chāndogya Upaniṣad

Hari-Bhakti-Vilāsa

Harivaṃśa Purāṇa

Kausitaki Brāhmaṇa

Mahābhārata

Manu Smṛiti (Laws of Manu)

Matsya Purāṇa

Naighaṇṭuka

Narsiṃha Purāṇa

Rājataraṁgiṇī

Rāmāyaṇa

Ṛg Veda

Sama Veda

Śatapatha Brāhmaṇa

Shiva Digvijaya

Skanda Purāṇa

Tevigga Sutta

Vasiśtha Dharma Śāstra

Vishnu Dharmottra

INDEX

Abhayāraṇya (forest reserve), 58
Ābhīras, 66
Abhiseka (ritual bathing of idol), 166
Aditi, 48, 54
Aga Khan, 80
Agarwal, 24, 34, 123, 141, 145, 160, 225
Agni, 4, 51
Agni Purāṇa, 166
Agra, 72, 147
Ahiṃsā: and food avoidances, 2–3; and hunting (*śikar*), 3; impact of, on wildlife, 3; preached by Mahāvīra, 3; and sacred cow concept, 6, 55; in Hinduism, 6, 56, 155, 156, 203–204; impact of, on Jain behavior, 17, 154, 203, 204; violated by killing of insects, 22; violated by cow slaughter, 24; origins of, 55–56; in Vedic literature, 55; in *Chāndogya Upaniṣad,* 55; in Jainism, 55, 58, 153–155; in *Bhagavad Gītā*, 56; in *Arthaśāstra*, 56; in Buddhism, 58, 256; in medieval India, 62; as state policy, 63; *pinjrapole* as expression of, 97, 102, 155, 157; derivation of word, 153–154; mentioned, 1, 6–8 passim, 16, 17, 22, 27, 53, 57, 60, 63, 69, 92, 102, 160, 163, 164, 171, 185, 198, 206
Āhīrs, 67, 237
Ahmad Shah (founder of Ahmedabad), 72, 80, 249
Ahmedabad: textile industry in, 72, 75, 87; as medieval trade center, 72; social and economic infrastructure of, in medieval period, 72; Moslems in, 72; Jains in, 72, 74, 78; *mahājans* in, 74–75; *Nagarseth* in, 74, 78, 252; animal homes in, 75; dispute over contributions to pinjrapole in, 76–78, 80, 167; Jain petition concerning pinjrapole tax in, 76–78; Vaishnavas in, 78; Vaishnava petition concerning pinjrapole tax, 80; mentioned, 19, 39, 71, 95, 143, 167, 252, 258
Ahmedabad District, 34
Ahmedabad Jain Swetambara Murti Pujak Visa Śrimali Friendly Society, 89, 251
Ahmedabad Pinjrapole (Ahmedabad Pinjrapole Society): animal population of, 16, 82–86; intake of animals during 1973 at, 16; *jīvat khān* at, 19, 22, 82; founding of, 75; early records of, 75–76; dispute over pinjrapole tax for, 76–78, 80; supported by tax on business transactions, 78; described in Bombay Gazetteer, 80–81; supported by Jains, 81, 89, 92; supported by *mahājans*, 81–82, 89, 91; composition of managing committee and advisory board of, 81, 250–251; effects of 1972–1974 drought on animal population of, 82, 87; *pārabaḍī* at, 82; cost of maintaining animals at, 87, 199, 201; mor-